The History of Astronomy: A Very Short Introduction

VERY SHORT INTRODUCTIONS are for anyone wanting a stimulating and accessible way in to a new subject. They are written by experts, and have been published in more than 25 languages worldwide.

The series began in 1995, and now represents a wide variety of topics in history, philosophy, religion, science, and the humanities. Over the next few years it will grow to a library of around 200 volumes – a Very Short Introduction to everything from ancient Egypt and Indian philosophy to conceptual art and cosmology.

Very Short Introductions available now:

Available soon:

For more information visit our web site

www.oup.co.uk/vsi

Michael Hoskin

THE HISTORY OF ASTRONOMY

A Very Short Introduction

OXFORD
UNIVERSITY PRESS

OXFORD

UNIVERSITY PRESS

Great Clarendon Street, Oxford OX2 6DP

Oxford University Press is a department of the University of Oxford.
It furthers the University's objective of excellence in research, scholarship,
and education by publishing worldwide in

Oxford New York

Auckland Bangkok Buenos Aires Cape Town Chennai
Dar es Salaam Delhi Hong Kong Istanbul Karachi Kolkata
Kuala Lumpur Madrid Melbourne Mexico City Mumbai Nairobi
São Paulo Shanghai Taipei Tokyo Toronto

Oxford is a registered trade mark of Oxford University Press
in the UK and in certain other countries

Published in the United States
by Oxford University Press Inc., New York

© Michael Hoskin 2003

British Library Cataloguing in Publication Data

Data available

Library of Congress Cataloging in Publication Data

Data available

ISBN 0-19-280306-9

1 3 5 7 9 10 8 6 4 2

Typeset by RefineCatch Ltd, Bungay, Suffolk
Printed in Spain by Book Print S. L., Barcelona

Contents

List of illustrations

The publisher and the author apologize for any errors or omissions in the above list. If contacted they will be pleased to rectify these at the earliest opportunity.

Chapter 1
The sky in prehistory

Historians of astronomy work mainly with the surviving documents from the past (fragmentary in quantity from antiquity, overwhelmingly bulky from recent times), and with artefacts such as instruments and observatory buildings. But can we discover something of the role the sky played in the 'cosmovision' of those who lived in Europe and the Middle East *before* the invention of writing? Could there even have been a prehistoric science of astronomy, perhaps one that enabled an elite to predict eclipses?

To answer these questions we rely primarily on the surviving stone monuments – their alignments, their relationships to the landscape, and the (usually ambiguous) carvings we find on some of them. The underlying problem of methodology is at its most acute when we are dealing with a monument that is unique. Stonehenge, for example, faces midsummer sunrise in one direction and midwinter sunset in the other. How can we be sure that an alignment that to us is of astronomical significance was chosen by Stonehenge's architects for this very reason? Did it have some quite different motivation, or even occur purely by chance? To take another example, a monument built around 3000 bc that faces east may have been oriented on the rising of the Pleiades, a bright cluster of stars in the constellation Taurus. It may have faced midway between midsummer and midwinter sunrise. Perhaps there was a sacred mountain in that direction. Or the orientation may have been chosen simply to take advantage of the slope of the ground. How can we decide which of these, if any, was in the minds of the builders?

We are on safer ground when we are dealing with a large number of monuments spread over a wide area. Archaeologists of western Europe study the communal tombs of the late Stone Age (the Neolithic), by which time the nomadic life of the hunter-gatherer had been replaced by the more settled existence of the farmer. Such tombs were to serve the needs of the clan for many years, and so they had an entrance through which additional bodies could be introduced as need arose. We can define the orientation of the tomb to be the line of sight of the bodies within as they 'look' out through the entrance.

In central Portugal there are many such tombs with a very characteristic and instantly recognizable shape, structures built by people with shared customs. They are scattered over a mountainless region that measures some 200 km from east to west and a similar distance from north to south; yet every single one of the 177 tombs that the writer has measured faces easterly, within the range of sunrise.

Not only that, but the directions in which the Sun rose in the autumn and winter predominate. Now we know from written records that Christian churches in many countries have traditionally been oriented to face sunrise (twice in the year), because the rising Sun is a symbol of Christ; and the builders often ensured this would happen by laying the church out to face sunrise on the very day that construction began. Suppose the Neolithic builders of these tombs followed a similar custom; suppose they too saw the rising Sun as a symbol of a life to come. Then, since it was doubtless in the autumn and winter, after the harvest, that they were free to dedicate themselves to such work, we would expect to find just such a pattern of orientations as we do in fact find; and it is difficult to imagine any other way of accounting for the striking

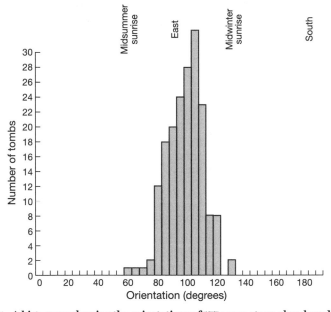

1. A histogram showing the orientations of 177 seven-stone-chambered tombs of central Portugal and adjacent regions of Spain. When the horizon altitudes have been taken into account, we find that every single tomb faced sunrise at some time of the year, most of them in the autumn months, when we might expect the constructors to have been at liberty to undertake such work. This is consistent with a custom of orienting tombs to face sunrise on the day construction began, as was later practised with Christian churches in England and elsewhere.

pattern of orientation. It seems reasonable, therefore, to infer that the Neolithic builders oriented their tombs on sunrise.

If this is so, then we have evidence that the sky played a role in the Neolithic cosmovision, just as it played (and plays) a role in the cosmovision of church builders. But this has nothing to do with 'science'. Claims that there was indeed an authentic science of astronomy in prehistoric Europe were made a generation ago by Alexander Thom, a retired engineer who took it upon himself to survey hundreds of stone circles in Britain. According to Thom, the prehistoric builders located the circles so that from them, the Sun (or Moon) could be seen rising (or setting) behind a distant mountain on a significant day – the winter solstice, for example, in the case of the Sun. For several days around the solstice, the Sun rises (or sets) at almost the same position on the horizon, and only with a very accurate instrument can the actual day of the solstice be identified. According to Thom, the prehistoric elite used circles in combination with distant mountains to form instruments many miles in extent; and, with the knowledge of solar and lunar cycles that this gave them, they could predict eclipses, and thus confirm their ascendancy over those around them.

Thom's work aroused enormous interest and, of course, controversy. However, a reinvestigation of his sites concluded that he had singled out those distant mountains that he knew would fit his ideas, and that such alignments could well have occurred by chance, and have been of no concern whatever to the prehistoric builders. Few now give credence to Thom's speculations, although anyone attempting to understand the cosmovision of a prehistoric people owes him a debt of gratitude for drawing attention to such issues.

We can be sure that in prehistoric times the sky served the practical needs of at least two groups: navigators and farmers. Today, in the Pacific and elsewhere, navigators use the Sun and stars to find their way, and no doubt prehistoric sailors in the Mediterranean did the same; but few if any traces of this survive. Concerning the agricultural calendar – farmers have always needed to know when to plant and when to harvest – we have some clues. Even today there are places in Europe where farmers avail themselves of celestial signals of the type described for us in *Works and Days* by the Greek poet Hesiod (*c.* 8th century BC). Each year the Sun completes a circuit of the stars, and there is therefore a period of some weeks when a given star – Sirius, for example – is too close to the Sun to be visible in the daytime. But the Sun moves on, and the day comes when Sirius can be glimpsed in the dawn sky: its 'heliacal rising'. Hesiod describes the sequence of heliacal events used by the farmers of his day for their calendar, and this must encapsulate the knowledge and experience assembled over the preceding centuries. Surprisingly, there seems to be much earlier evidence of just such a sequence inscribed on pillars of the temple of Mnajdra on Malta, which dates from around 3000 BC. My colleagues and I found rows of incised holes that seem to be tallies, and on analysing the counts we found that they may well indicate the numbers of days between one important heliacal rising and the next. As we shall see, the heliacal rising of Sirius was soon to play a pivotal role in the calendar of nearby Egypt.

Chapter 2
Astronomy in antiquity

The origins of modern astronomy first emerged from the mists of prehistory in the 3rd and 2nd millennia before Christ, in the increasingly complex cultures that developed in Egypt and Babylon. In Egypt the effective administration of a far-flung kingdom depended upon a well-established calendar, while rituals called for the ability to tell the time at night, and for the capacity to orientate monuments – pyramids – in the cardinal directions. In Babylon the security of the throne, and therefore of the state, depended upon the correct reading of omens, including those seen in the sky.

Calendars were, and are, awkward to formulate because there is no exact number of days in either the lunar month or the solar year, and likewise no exact number of months in the year; our own extraordinary jumble of month-lengths is a symptom of the problems nature poses for the calendar maker. In Egypt life was dominated by the annual flooding of the Nile, and a solution to the calendar problem was found when it was noticed that this flooding took place around the day when Sirius rose heliacally – when the star appeared in the dawn sky after an absence of weeks. The star's rising could therefore be used to anchor the calendar.

Each year consists of 12 lunar months and about 11 days, and the Egyptians devised a calendar in which Sirius would *always* rise in the 12th month. If in any given year the star rose early in the

12th month, well and good: it would rise next year in the same month. But if it rose late in the 12th month, then unless action was taken, the following year it would rise after the month had ended. To prevent this happening, an extra, or 'intercalary', month would be declared for the current year.

Such a calendar was satisfactory for religious festivals, but not for the administration of a complex and highly organized society, and so for civil purposes a second calendar was devised. It was ruthlessly simple: every year consisted of exactly 12 months, each of three 'weeks' of ten days, together with an extra five days at the end of the year to bring the total number of days to 365. As the seasonal year is in fact a few hours longer (which is why we have leap years), this civil calendar slowly cycled through the seasons; but this was considered an acceptable price to pay for the administrative convenience of an unchanging pattern.

Since there were 36 'weeks' of ten days, 36 star groups or 'decans' were selected around the sky so that a new decan rose heliacally every ten days or so. As dusk fell on any given night a number of decans would be visible overhead, and during the night new ones would appear on the horizon at regular intervals, so marking the passage of time.

The sky played a profound role in Egyptian religion, for deities were present there in the form of constellations, and immense labour was expended on Earth to ensure that the reigning pharaoh would one day join them. We see one aspect of this in the almost precise north–south alignments of the funerary pyramids of pharaohs of the 3rd millennium, and there has been much debate as to how this was achieved. A clue comes from the (tiny) errors in the alignments, for these errors change systematically with the dates of construction. It has recently been suggested that the Egyptians may have referred to an imaginary line joining two particular stars that were seen above the horizon at all times (circumpolar stars), and have taken north to be the direction towards this line at the exact moment when the line was vertical. If so, the slow movement of the celestial north pole due to the wobble of the Earth's axis (called precession) would account for the systematic errors.

The Egyptians were handicapped by the primitive condition of their geometry and arithmetic, and this precluded them from developing an understanding of the more subtle movements of the stars and planets. In particular, their arithmetic operated almost exclusively with fractions that had the number one in the numerator.

By contrast, 2,000 years before Christ the Babylonians developed a brilliant technique for arithmetical notation, and this was the basis of their remarkable achievement in astronomy. A scribe would take a soft clay tablet the size of a man's hand, and impress his stylus on it edgeways to denote a 1, and flatways to denote 10. Doing this as often as necessary, he would write numbers from 1 to 59; but for 60 he would use the symbol for 1, much as we do in writing 10, and similarly for 60×60, $60 \times 60 \times 60$, and so on. There were no limits to the accuracy and versatility of the numbers that could be written in this sexagesimal system of notation, and even today we continue to write angles in sexagesimal degrees, and to reckon time in hours, minutes, and seconds.

Babylonian court officials were on the alert for omens of all kinds – the entrails of sheep were of special interest – and they kept records of any unwelcome events that ensued, so as to learn from experience: when the omen occurred again in the future, they would know the nature of the impending disaster of which the omen was a warning, and so the appropriate ritual could be performed. This led to the compilation of a vast compendium of 7,000 omens that had taken definitive form by 900 BC.

Soon thereafter the scribes began systematically to record astronomical (and meteorological) phenomena, in order to refine their prognostications. For seven centuries this continued, and gradually cycles in the movements of the Sun, Moon, and planets

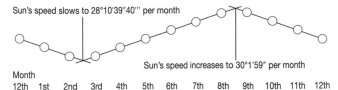

Sun's speed slows to 28°10′39″40‴ per month

Sun's speed increases to 30°1′59″ per month

Month
12th 1st 2nd 3rd 4th 5th 6th 7th 8th 9th 10th 11th 12th

2. A repesentation in modern terms, with the values found in a tablet for 133/132 BC, of the second Babylonian approximation of the speed of the Sun against the background stars. In this artificial but arithmetically convenient formulation, the speed is imagined to increase by the same amount each month for six months, and then to decrease similarly for the next six months. This was found to yield acceptably accurate results.

began to emerge from the records. With the help of their sexagesimal notation, the scribes devised arithmetical techniques for using these cycles to predict the future positions of the celestial bodies. For example, the Sun's movement against the background stars accelerates for one half of the year, and decelerates for the other half. The Babylonians devised two techniques for approximating to this movement: either they assumed one uniform speed for half the year and another uniform speed for the remaining half; or they assumed a steady increase in speed for half the year and a steady decrease for the remaining half. Both were no more than artificial approximations to reality, but they did the job.

Of Greek astronomy prior to the 4th century BC our knowledge is fragmentary in the extreme, for few writings survive from this period, and much of what we have is in the form of citations by Aristotle (384–322 BC) of opinions he is about to attack. Two aspects, however, stand out: first, the emergence of an attempt to understand nature in purely natural terms, without recourse to the supernatural; and second, the recognition that the Earth is a sphere. Aristotle rightly points out that the shadow of the Earth cast on the Moon during an eclipse is invariably circular, and that only if the Earth is a sphere can this be the case.

Not only did the Greeks know the shape of the Earth, but

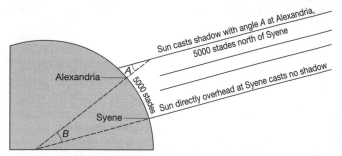

3. The geometry used by Eratosthenes to measure the Earth. Angles *A* and *B* are equal.

Eratosthenes's measurement of the circumference of the spherical Earth

Eratosthenes believed that at what is now Aswan, the Sun was overhead at noon on midsummer's day, whereas at Alexandria, thought to be 5,000 stades due north of Aswan, it was one-fiftieth of a circle from being directly overhead. This being so, simple geometry showed that the circumference of the Earth was 50 times 5,000 stades. The modern equivalent of the stade is debated, but there is no doubt that the value of 250,000 stades was approximately correct.

Eratosthenes (*c*.276–*c*.195 BC) arrived at an excellent estimate of the Earth's actual size. Ever since then, everyone with a modicum of education has known that the Earth is spherical.

So, it seemed, was the sky. Furthermore, we always see exactly half the celestial sphere, and therefore the Earth must be at its very centre. And so developed the classic Greek model of the universe: a spherical Earth at the centre of a spherical cosmos.

Aristotle, in voluminous writings that were still being taught in

Cambridge in Isaac Newton's day, contrasted the terrestrial region at the centre of the cosmos – extending almost as far as the Moon – with the celestial region that lay beyond. In the terrestrial region there was change, life and death, coming to be and passing away. At the very centre was the sphere of Earth; around it, the shell of water, then the shell of air, and finally the shell of fire. Bodies were made of these elements in varying proportions. Left to itself, a body would move in a straight line, either towards the centre or away from it, in order to reach the distance from the centre appropriate to its elemental makeup: thus stones, being primarily earthy, fell down towards the centre, whereas flames rose towards the sphere of fire.

Immediately beyond the sphere of fire was the beginning of the celestial region, where the movements were cyclic (never rectilinear), and therefore there was no true change. Highest in the sky was the rotating sphere of the innumerable 'fixed' stars, so-called because they never altered their positions relative to each other. The stars that were not fixed numbered just seven: the Moon (clearly the nearest of all), the Sun, Mercury, Venus, Mars, Jupiter, and Saturn. These moved against the background of the fixed stars, and because their movements were forever changing – indeed, the five lesser bodies actually reversed direction from time to time – they were known as 'wanderers', or 'planets'. Aristotle's teacher Plato (427–348/7 BC), a mathematician by instinct, had seen the planets as possible disproof of his belief that we live in a cosmos governed by law: but could it perhaps be shown that the planets were in fact as regular in their movements as the fixed stars, the sole difference being that the laws governing the planetary movements were more complex, and not immediately obvious?

The challenge was taken up by the geometer Eudoxus (c.400–c.347 BC), who formulated for each planet a nest of either three or four concentric spheres, which he used in a mathematical demonstration that the planets' movements were lawlike after all. Each planet was imagined as being on the equator of the innermost sphere, which rotated with uniform speed, carrying the planet with

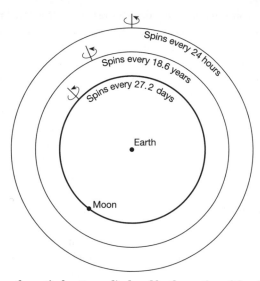

4. The mathematical patterns displayed by the motion of the planet Moon, according to Eudoxus. The Moon is imagined to be on the equator of the innermost sphere, which rotates once a month. The poles of this sphere are embedded in the next sphere, which rotates every 18.6 years, a period familiar from eclipse cycles; and the poles of this sphere are embedded in the outermost sphere, which rotates daily.

it. The poles of this sphere were thought of as embedded in the next sphere and carried round by it as it too uniformly rotated, and so on for the third and (in the case of the lesser planets) the fourth. The angle of the axis of each sphere was carefully chosen, as was its speed of rotation, with the outermost sphere in each case generating the daily path of the planet around the Earth. The spheres of the Moon, for example, rotated with uniform speed every 24 hours, every 18.6 years, and every 27.2 days, respectively, and so the resultant motion of the Moon reflected all three periods.

For each of the five lesser planets, two of the spheres rotated with equal and opposite speeds about axes that differed only slightly, and these spheres by themselves would give the planet a motion in a figure-of-eight; this allowed the complete nests of four spheres to generate backward motions from time to time.

So far, so good. But in these geometric models the backward (retrograde) motions of the lesser planets repeated themselves with complete regularity, and clearly did not reproduce the erratic movements of the planets that we actually observe in the sky. Furthermore, the models forced each planet to remain at a constant distance from the central Earth, whereas in the real world the lesser planets vary considerably in brightness and so presumably in distance from us. Such shortcomings would have been anathema to a Babylonian, but the models were sufficiently promising to satisfy Plato's generation that the cosmos was indeed lawlike, even if its laws had yet to be elucidated completely.

Aristotle was exercised about a quite different limitation: the spheres of the models were constructions in the minds of mathematicians, and did not explain in physical terms how the planets come to move as we observe them to do. His solution was to convert the mathematical spheres into physical reality, and to combine them to make one composite nest for the entire system. The daily rotation of the outermost sphere of all, that of the fixed stars, now sufficed to impose a daily rotation on every planet within, so the outermost sphere of each planetary nest could be discarded. However, the spheres special to an individual planet would transmit their motion down through the system, unless steps were taken to prevent this; and Aristotle therefore interpolated additional spheres with opposite motions in the appropriate places, to cancel out any unwanted rotations.

The resulting Aristotelian cosmology – a central terrestrial or sublunary region where there was coming to be and passing away, and an outer celestial region whose spheres generated the cyclic movements of the fixed stars and planets – was to dominate Greek, Islamic, and Latin thought for the better part of two millennia.

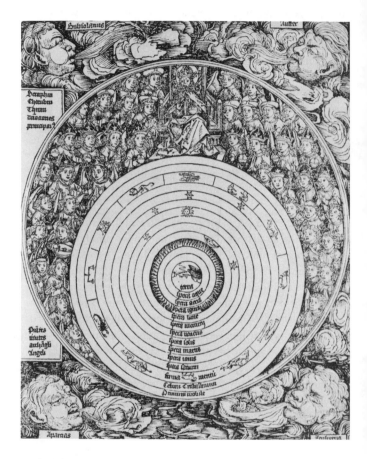

5. Aristotle's cosmos in Christian guise, as portrayed in the *Nuremberg Chronicle* of 1493. At the centre are the four elements (earth, water, air, and fire); then come the spheres of the planets (Moon, Mercury, Venus, Sun, Mars, Jupiter, and Saturn), followed by the spheres of the firmament of stars, the crystalline heaven, and the First Mover. Outside we see God enthroned with the nine orders of angels.

Yet its inflexibility and the resulting gap between theory and observation gained significance almost immediately, as Aristotle's pupil, Alexander the Great, conquered much of the known world and Greek geometrical astronomy began to merge with the arithmetical and observation-based astronomy of the Babylonians. Uniform circular motions continued to be seen by Greek astronomers as the key to understanding the universe, but now they were to be employed with more flexibility and with greater concern for observational fact.

Around 200 BC the geometer Apollonius of Perga developed two geometrical tools that supplied this flexibility. In one, the planet moved uniformly on a circle, but the circle was now *eccentric* to the Earth.

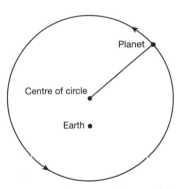

6. In an eccentric circle, the planet moved as usual in a circle around the Earth with uniform speed; but because the Earth was not at the centre of the circle, the planet's speed appeared to vary.

As a result, the planet would appear to move faster when its path brought it nearer the Earth, and slower when it was away on the far side of its orbit. In the other, the planet was located on a little circle, or *epicycle*, whose centre was carried around the Earth on a *deferent* circle.

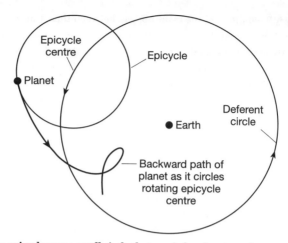

Epicycle
centre

Epicycle

Planet

Deferent
circle

Earth

Backward path of
planet as it circles
rotating epicycle
centre

7. An epicycle was a small circle that carried a planet moving around the circle with uniform speed. The centre of the circle likewise moved with uniform speed on a deferent circle around the Earth. The diagram indicates how the apparent backwards motion observed from time to time in the five lesser planets could be imitated in the model if the two speeds were appropriately chosen.

It is easy for us to appreciate the value of this device, because – as seen by us – Venus (for example) orbits the Sun, which in turn orbits the Earth. Astronomy, one might say, was on the right track; not only that, but on a most promising one, for repeated refinement of the various quantities (parameters) involved would lead to encouraging progress, but never to total success – until at last Kepler abandoned circles in favour of ellipses.

The first to employ these devices was Hipparchus, who made observations at Rhodes between 141 and 127 BC. Although all but one of his works are lost, having become obsolete when later subsumed into the *Almagest* of Ptolemy, we are informed about his achievements by what we read of him in the *Almagest*. It was through Hipparchus that the geometrical astronomy of the Greeks began to incorporate precise parameters derived from the long centuries during which the Babylonians had kept their observational records. Hipparchus compiled a list of lunar eclipses observed at Babylon from the 8th century BC, and these records were crucial to his study of the motions of the Sun and Moon, for it is during an eclipse that these two bodies are exactly in line with the Earth. Hipparchus adopted the Babylonian sexagesimal system for writing numbers, and divided the ecliptic and other circles into 360 degrees. He succeeded in reproducing the solar motion by means of a single eccentric circle, and Ptolemy was to take over this model almost unchanged. He was less successful with the Moon, while the lesser planets he left to his successors.

Hipparchus's single most important discovery was that of the precession of the equinoxes, the slow movement among the stars of the two opposite places where the Sun crosses the celestial equator. The spring equinoctial point is used by astronomers to define their frame of reference, and the movement of this point implies that the measured position of a star varies with the date of measurement.

Hipparchus also compiled a star catalogue, but this is lost; the only surviving catalogue from antiquity is the one in the *Almagest*. Whole forests have been sacrificed to the debate among historians as to whether Ptolemy himself observed the positions given in his catalogue, or whether he took the positions as observed by Hipparchus and simply converted them to his own epoch by correcting for precession.

The three centuries that separated Hipparchus and Ptolemy were a dark age for astronomy; at least, Ptolemy seems to have despised whatever was done in that time and tells us little about it. Most of what information we have has been recovered from later writings in Sanskrit, for Indian astronomy was very conservative and its writers preserved what they had learned from the Greeks. But when we come to the *Almagest* itself, we are on safer ground. Of the author's life we know little, but he reports observations he made between AD 127 and 141 in the great cultural centre of Alexandria, and so he cannot have been born much later than the beginning of the 2nd century. He may well have spent his adult life in Alexandria, home of the great museum and library, and he is (like Hipparchus) an example of a Greek astronomer who flourished in a place remote from the Greek mainland but close to the irreplaceable observational records of the Babylonians.

The *Almagest* is a magisterial work that provided geometrical models and related tables by which the movements of the Sun, Moon, and the five lesser planets could be calculated for the indefinite future. Written half a millennium after Aristotle, when Greek civilization had almost run its course, it synthesizes the Graeco-Babylonian achievement in mastering the movements of the wanderers. Its catalogue contains over 1,000 fixed stars arranged in 48 constellations, giving the longitude, latitude, and apparent brightness of each. As the works of earlier authors, notably Hipparchus, were now obsolete, they vanished from the face of the Earth, and the *Almagest* would dominate astronomy like a colossus for 14 centuries to come.

But there were problems ahead. Aristotelian cosmology had explained the heavens in terms of spheres concentric with the Earth, and philosophers felt comfortable with such spheres and their uniform rotations. However, Apollonius and Hipparchus had introduced eccentres and epicycles that violated this convention; true, the planets in such models moved uniformly on their circles, but not about the Earth. This was bad enough, but Ptolemy found it necessary to use a still more questionable device in order to 'save the appearances' of the planetary motions with economy and accuracy: the *equant point*.

In a planetary model, the equant was the point symmetrically opposite the eccentric Earth, and the planet was required to move on its circle so that from the equant point it would *appear* to be moving uniformly across the sky. But since the equant point was *not* at the centre of the circle, to do this the planet would have to vary its

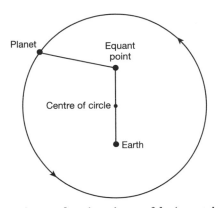

8. The equant point was the mirror image of the (eccentric) Earth, and the planet was supposed to move in such a way that from the equant point it would appear to be moving uniformly. In fact, therefore, the planet moved non-uniformly.

speed. Ptolemy was an astrologer anxious to know the positions of the planets at all times (his *Tetrabiblos* is a classic of astrology); and accurate prediction – however questionable the means used to achieve it – had higher priority than fidelity to the philosophical norm that all motions on circles be uniform. For him, as for the Babylonians, accuracy, not truth, was the primary consideration.

Kepler's laws of planetary motion reveal to us just why the equant was such a useful geometrical tool.

In conformity with the first two laws, the Earth (for example) moves in its orbit around the Sun in an ellipse, with the Sun at one of the two foci; and the line from the Sun to the Earth traces out equal areas in equal times. Therefore, when the Earth in its orbit is near the Sun it moves faster than usual, and when it is far from the Sun (and therefore near the other, 'empty' focus of the ellipse) it moves more slowly. Viewed from the empty focus, the Earth's speed across the sky will appear almost uniform: for when near the Sun and far from the empty focus the Earth is moving faster than usual, and this is masked by its greater distance from the empty focus; when near

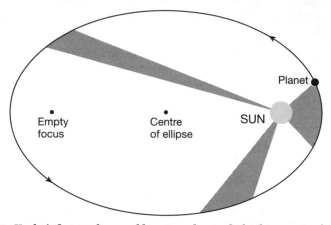

9. Kepler's first two laws enable us to understand why the equant point was a useful tool. They imply that a planet orbiting the Sun in an ellipse moves faster when near the Sun and slower when near the other focus of the ellipse. As a result, when viewed from this 'empty' focus, the planet's movement will appear approximately uniform. In this diagram the ellipticity of the orbit has been greatly exaggerated.

the empty focus the Earth is moving more slowly than usual, and this is masked by its nearness to the empty focus. In other words, Kepler teaches us that to an approximation, the Earth's speed across the sky as viewed from the empty focus is indeed almost uniform; and the empty focus is the counterpart in the Keplerian ellipse of the equant in the Ptolemaic circle.

In the universities of the later Middle Ages, students would be taught Aristotle in philosophy and a simplified Ptolemy in astronomy. From Aristotle they would learn the basic truth that the heavens rotate uniformly about the central Earth. From the simplified Ptolemy they would learn of epicycles and eccentrics that violated this basic truth by generating orbits whose centre was not the Earth; and those expert enough to penetrate deeper into the Ptolemaic models would encounter equant theories that violated the (yet more basic) truth that heavenly motion is uniform. Copernicus, among many others, would be shocked.

Nevertheless, with the models of the *Almagest* – whose parameters would be refined over the centuries to come – the astronomer, and the astrologer, could compute the future positions of the planets with economy and reasonable accuracy. There were anomalies – the Moon, for example, would vary its apparent size dramatically in the Ptolemaic model but does not do so in reality, and Venus and Mercury were kept close to the Sun in the sky by a crude *ad hoc* device – but as a geometrical compendium of how to grind out planetary tables, the *Almagest* worked, and that was what mattered.

In his *Planetary Hypotheses*, composed after the *Almagest*, Ptolemy presented his cosmology. Like earlier Greek cosmologists, he thought it reasonable to assume that the longer a planet takes to orbit the heavens against the background of the stars – that is, the less difference there is between its motion and the regular daily motion of the stars – the nearer it is to the stars. This being so, Saturn, with its period of 30 years, was nearest to the fixed stars and furthest from Earth, followed by Jupiter (12 years) and Mars (two years) in descending order. The Moon (one month) was nearest to Earth. But what of the Sun, Venus, and Mercury, which keep company as they move among the stars and therefore all have the same period of one year? Because of the Sun's dominance in the sky, and because some planets kept it company while others did not, the Sun was traditionally taken to be the middle one of the seven planets, located immediately below Mars and separating the planets that kept it company from those that did not. The order of Venus and Mercury had long been disputed; Ptolemy placed Mercury below Venus, on little more than the toss of a coin.

Having established the order of the planets – by reasoning that varied from the plausible to pure guesswork – Ptolemy now made the assumption that every possible height above the Earth was from time to time occupied by one particular planet, and only one. Thus the greatest height of the Moon (Ptolemy had an argument to show that this was 64 Earth radii) equalled the least height of the next planet, Mercury. The geometrical model for Mercury specified the ratio between its least and greatest heights, and multiplying 64 Earth radii by this ratio gave the greatest height of Mercury; and so on. That is, the geometrical models gave the ratios between the least and greatest heights of each successive planet, and the maximum lunar height of 64 Earth radii calibrated the entire system. The fixed stars, which lay at the maximum height of Saturn, were 19,865 Earth radii above us, or some 75 million miles: the Ptolemaic universe was impressively large.

Hipparchus had begun the process of employing the tools provided by the Babylonian tradition – arithmetical versatility and the use of long centuries of observation to determine parameters of exceptional accuracy – to pursue the central ambition of Greek geometrical astronomy: to reproduce the entire orbit of each planet by means of a geometrical model based on the fundamental cosmological principle of uniform circular motion. Ptolemy had carried the process to fruition, albeit with some compromises. The models of the *Almagest* would be refined time and again in the years to come; but only after 14 centuries, and the invention of printing, would a mathematical astronomer of equal competence consider its defects so fundamental as to call for a reformation.

Chapter 3
Astronomy in the Middle Ages

It was in AD 622 that the prophet Mohammad fled Mecca to Medina, and before long the new religion of Islam had spread across the whole of North Africa and into Spain. Islam made specific demands upon the skills of astronomers. The month began with the new moon – not when Sun, Moon, and Earth were geometrically aligned, but two or three days later, when the crescent was seen by human eyes. Could this be regularized, so that neighbouring villages would agree on the beginning of the new month, even when the sky was clouded? The hours of prayer were set by the altitude of the Sun as it traversed the sky, and the need to determine these hours correctly eventually led to the institution of the office of *muwaqqit*, or mosque timekeeper, so giving astronomers a secure and respected position in the community. And the determination of the local direction of Mecca, the *qibla*, which dictated the layout of mosques and graveyards and much else, posed a challenging problem that *muwaqqits* and other astronomers sought to solve.

Long before the arrival of Islam, the great centre of learning in Alexandria had fallen on troubled times. The *Almagest* itself was to find its way to Constantinople, and in the 9th century a copy was purchased by emissaries from Baghdad, where the youthful and vibrant Muslim culture had woken up to the intellectual treasures surviving in the Greek language. At

Baghdad it was translated by a team working in the House of Wisdom, first from Greek into Syriac and then from Syriac into Arabic. Other copies in Constantinople would gather dust, unread, until in the 12th century the emperor presented one as a ceremonial gift to the King of Sicily, where it was translated into Latin.

Despite the censure in the Koran, astrology flourished in the Muslim world at every level of society; and those astrologers who were more than mere fortune tellers based their predictions on tables of planetary positions. The success of the models of the *Almagest* was undisputed, but these models incorporated parameters that could be determined with ever greater accuracy as the centuries passed – and Ptolemy himself had explained how to do this. At first the measuring instruments used by astronomers for this purpose were modest in size, but as the observers' ambitions grew so did the size of their instruments, and they looked to patrons to pay for their construction and housing.

At times, however, this aroused the hostility of the religious authorities, and a patron's death – or even his loss of nerve – could bring astronomical observation to an end. In Cairo the construction of an observatory began in 1120 on the order of the vizier, but in 1125 his successor was killed by command of the caliph, his crimes included communication with Saturn, and the observatory was demolished. In Istanbul an observatory for the astronomer Taqi al-Din was completed by Sultan Murad III in 1577 – as it happened, just in time for observations of a bright comet. Taqi al-Din, doubtless with an eye to his own prosperity, interpreted the apparition as boding well for the sultan in his fight against the Persians. But events turned out otherwise, and in 1580 religious leaders convinced the sultan that it was inviting misfortune to pry into the secrets of nature. The sultan therefore ordered the observatory to be destroyed 'from its apogee to its perigee'.

Only two Islamic observatories enjoyed more than a brief existence. At Maragha, the present-day Maragheh in northern Iran, construction of an observatory for the distinguished Persian astronomer Nasir al-Din al-Tusi (1201–74) was begun in 1259 by Hulagu, the Mongol ruler of Persia. Its instruments included a 14-foot-radius mural quadrant (an instrument for measuring altitudes, attached to a wall aligned in the north–south direction) and an armillary sphere (used for other measurements of position) with circles five feet in radius. With the help of these instruments, a team of astronomers completed in 1271 a *zij*, or collection of astronomical tables with instructions for their use, in the tradition of Ptolemy's own *Handy Tables*. But in 1274 al-Tusi left Maragha for Baghdad, and although observations at the observatory continued into the next century, its creative period was already over.

The other major observatory of Islam enjoyed the advantage that the prince himself was an enthusiastic member of staff. At Samarkand in central Asia, Ulugh Beg (1394–1449), a provincial governor who was to succeed to the throne in 1447, began construction of a three-storey building in 1420. Its chief instrument, built on the principle that 'bigger is better', was a form of sextant no less than 130 feet in radius. This was mounted out of doors between marble walls aligned north–south, and the range of the instrument was chosen so that it could be used to observe the transit of the Sun, Moon, and the other five planets. The great achievement of Samarkand Observatory was a set of astronomical tables that included a catalogue of over 1,000 stars. Much earlier, the Baghdad astronomer Abd al-Rahman al-Sufi (903–86) had prepared a revision of Ptolemy's star catalogue in which he gave improved magnitudes and Arabic versions of the identifications; but he had left the stars themselves and their often inaccurate relative positions unchanged, and so Ulugh Beg's was to be the single important star catalogue of the Middle Ages. Samarkand Observatory fell into disuse soon after the murder of Ulugh Beg in 1449.

Observatories were for an elite, but every astrologer needed to make observations, and this became possible through the development of the astrolabe, an ingenious portable computer and observing instrument that had its roots in antiquity. The typical astrolabe consisted of a brass disc that could be suspended by use of a ring at the top edge. One side of the astrolabe was for observations of the angular altitude above the horizontal of a star or planet; the observer suspended the instrument and looked at the heavenly body along a sighting bar, and then read the angle on a scale around the circumference. The other side of the disc represented the celestial sphere projected on to the plane of the equator from the south celestial pole.

10. A 14th-century astrolabe preserved at Merton College, Oxford

Each line from the pole intersected the celestial sphere in one further point, and it intersected the plane of the equator (in a single point); the latter point was the 'projection' of the former. Because the brass disc was of course of finite size, and because the heavens south of the Tropic of Capricorn were of no practical interest, the projected skies extended from the north celestial pole (represented by the centre of the disc) as far as this Tropic but not beyond.

Circles of equal altitude at the latitude of the observer projected into cirles that were engraved on the disc, along with much else. So far, so good; but the stars of the rotating heavens also needed representation. This was achieved by means of a brass sheet that bore indications of the locations of the principal stars but was otherwise cut away as much as possible to reveal the coordinate circles below. This sheet rotated about the central point of the disc underneath, just as the stars rotate about the north celestial pole. The sheet also contained a representation of the ecliptic path of the Sun, and the observer needed to know (and mark) the Sun's current position on it.

This done, a single observation – typically, of the altitude of a star at night or of the Sun by day – would allow the observer to rotate the sheet into its correct current position, by moving it until the star (or the Sun) lay over the appropriate coordinate circle. This done, the entire heavenly sphere was now in position, and many questions could be answered – for example, which stars were currently above the horizon and what was the altitude of each. Time could be determined by aligning the Sun with a scale on the perimeter of the disc and reading off the hour on the scale. This was possible irrespective of whether it was by observation of the Sun or of a star that the sheet had been positioned: the astrolabe was a clock that could be used to tell the time night and day throughout the 24 hours.

A wide range of other information could be obtained easily from the astrolabe. For example, to determine the hour at which a given star would rise, the astronomer would rotate the sheet until the star was above the circle of zero altitude, and then read off the time. The astrolabe was a simple, ingenious, and versatile device that encouraged quantitative observation of the heavens.

A *zij* had been composed as early as the first half of the 9th century, in the House of Wisdom at Baghdad, by al-Khwarizmi, a corruption of whose name gives us the word 'algorithm'. It made use of the parameters and computational procedures contained in a Sanskrit astronomical work that had been brought there around 770. In a later version, the *zij* was to be translated into Latin in the 12th century, and so became a vehicle by which Indian astronomical methods reached the West. By making possible the prediction of future planetary positions, *zijes* supplied the needs of the practising astronomer/astrologer, and great numbers of these tables were produced, often using parameters that improved on those of Ptolemy.

Islam had no counterpart to the emerging universities of the Christian West, and we look in vain for an Islamic thinker of sufficient originality to challenge the foundations of Aristotelian/ Ptolemaic cosmology. Nevertheless, discussions of *shukuk*, or doubts, concerning Ptolemy were appearing regularly by the 10th century. The most obvious target was the Ptolemaic equant, which violated the basic principle of uniform circular motion, but the epicycle and eccentre also came in for criticism because they involved motions that, while uniform, did not take place about the central Earth. A philosophical purist in this was the Andalusian Muhammad ibn Rushd (1126–98), known to the Latins as Averroes; in the West Aristotle was to become 'The Philosopher', and Averroes 'The Commentator'. Averroes recognized that Ptolemaic models 'saved the appearances' – reproduced the observed motions of the planets – but this did not make them true. His contemporary and fellow Andalusian Abu Ishaq al-Bitruji (Alpetragius) attempted to devise alternative models that met Aristotelian requirements, but of course with very unsatisfactory results.

In Cairo Ibn al-Haytham (Alhazen, 965–c.1040) tried to adapt the Ptolemaic models so that they could take on physical reality. In his *On the Configuration of the World* the heavens were formed of concentric spherical shells, within whose thicknesses smaller shells and spheres were located. His work was translated into Latin in the 13th century, and was to become one of the influences on Georg Peurbach in the 15th century.

The equant had long aroused misgivings among even the most practical-minded astronomers, and at Maragha in the 13th century al-Tusi succeeded in devising a geometrical substitute involving two small epicycles; Copernicus was at one stage in his career to adopt a similar device, and for the same reason, though historians have not yet identified an unambiguous link between them. An attempt to devise planetary models that were purged of all objectionable elements was made by Ibn al-Shatir, *muwaqqit* of the Umayyad mosque at Damascus, in the middle of the 14th century. His lunar model avoided the huge variations in the apparent size of the Moon implied by the lunar model of the *Almagest*, his solar model was based on new observations, and all his models were free not only of equants but also of eccentres. Epicycles, however, he found unavoidable, for reasons that we can well understand. By the time of al-Shatir, however, the Latin West had developed its own astronomical tradition, and was no longer reliant upon translation from the Arabic.

This independence had been slow in coming. In the Roman world, Greek had continued to be the language of scholars, and none of the major astronomical works of antiquity was written in Latin. With the collapse of the Roman Empire, knowledge of Greek almost completely disappeared in the West, so that the classics of ancient astronomy – even if available – could no longer be read. Ancius Manilius Severinus Boethius (c.480–524/5), a high official in the Roman Gothic kingdom, set himself to translate into Latin as many treatises of Plato and Aristotle as possible, but he had already left it too late. However, before his execution for defiance of his king over an injustice, Boethius did manage to translate a number of Greek works, several of them on logic, and these he set alongside logical writings by Roman authors such as Cicero. In this way he bequeathed to later centuries a corpus of texts in what became the one secular area of study where the medieval student might 'compare and contrast' and so come to his own conclusions. As a result, logical consistency was to become an obsession in the medieval university, where debates over the validity of epicycles, or whether certainty could in principle be attained in a planetary model, were meat and drink to the young students in Arts.

Just one (incomplete) work of Plato made its way into Latin during this period: his cosmological myth, *Timaeus*, two-thirds of which was translated by Calcidius (in the 4th or 5th century), who supplied a lengthy commentary. Astronomical works written in Latin in the early Middle Ages make sad reading, although the basic fact of the sphericity of the Earth was never lost to sight. Ambrosius Theodosius Macrobius, an African who lived in the early 5th century, wrote a commentary on Cicero's *Dream of Scipio*, and in this he expounded a cosmology in which a spherical Earth lay at the centre of the sphere of stars, which rotated daily from east to west. As it did so, it dragged the planetary spheres with it, though each of these also had its individual motion in the opposite direction. Macrobius is vague about the order of the planets because his sources differed. Martianus Capella of Carthage (*c.*365–440) wrote *The Nuptials of Philology and Mercury*, an allegory of a heavenly marriage at which each of the seven bridesmaids presented a compendium of one of the Liberal Arts. This account of astronomy is notable for the explanation of why Venus and Mercury are always seen near the Sun in the sky: they are circling the Sun, and so they accompany the Sun as it circles the Earth.

Christianity, like Islam, presented challenges to astronomers, chief among them the calculation of the date of Easter. In simple terms, Easter Day is the Sunday that follows the full moon that follows the spring equinox, and so its date in any given year depends on the cycles of both Sun and Moon. It might have been possible for the Christians of Alexandria, as inheritors of the accurate values for month and year handed down from Babylon, to calculate the appropriate date of Easter for some years ahead; but the Church authorities took the more practical course of trying to identify a period consisting of a number of years that almost equalled an integral number of months, and establishing the dates of Easter within the approaching years of this period. Once established, such a sequence could be repeated for future periods, indefinitely.

The cycle eventually adopted was one discovered by Babylonian astronomers in the 5th century BC but credited to the Greek Meton, whereby 235 lunar months equal 19 years (with an error of only a couple of hours). The definitive treatise, *On the Divisions of Time*, was written in 725 by the Venerable Bede (672/673–735) of Jarrow in England. In the calendar laid down by Julius Caesar, a leap year occurred every fourth year (without exception); every four years, therefore, the day of the week on which a given date occurred advanced by five, and so in $7 \times 4 = 28$ years it would return to its original day. Bede combined this with the 19-year Metonic cycle to produce an overall cycle of $19 \times 28 = 532$ years that catered for the luni-solar pattern of Easter together with the requirement that it occur on a Sunday.

The revival of astronomy – and astrology – among the Latins was stimulated around the end of the first millennium when the astrolabe entered the West from Islamic Spain. Astrology in those days had a rational basis rooted in the Aristotelian analogy between the microcosm – the individual living body – and the macrocosm, the cosmos as a whole. Medical students were taught how to track the planets, so that they would know when the time was favourable for treating the corresponding organs in their patients.

In 1085 the great Muslim centre of Toledo fell into Christian hands, and the intellectual riches of Islam and, more especially, Greece became accessible. Translators descended on Spain, the most notable being Gerard of Cremona (*c.*1114–87), whose innumerable translations included the *Almagest* and the Toledan Tables of al-Zarqali (d. 1100). These tables were then adapted for other places and proved immensely successful, although the underlying planetary models remained a mystery for the time being.

If the 12th century was the era of translations, the 13th was that of the assimilation of the works translated. In the emerging universities, Latin was the *lingua franca*, and so there was no language barrier to prevent students and teachers going where they wished. Prospective lawyers might go to Bologna and medical students to Padua, but in most disciplines Paris was pre-eminent.

There, as elsewhere, the Faculty of Arts provided the basic education in literacy and numeracy, through the medium of the seven Liberal Arts, which included astronomy. The Arts students were mostly boys in their teens, and the invention of printing lay in the future, so the level of instruction was inevitably elementary. A minority of students would eventually stay on for theological, medical, or legal studies in one of the higher faculties; medicine and law enjoyed their traditional prestige, while the writings of Augustine and the other Fathers of the Church ensured that theology was a challenging intellectual discipline. There was therefore tension between the teachers in these higher faculties and those trapped in the humdrum routine of Arts.

The bulk of the new translations, however, belonged to Arts, and provided the Parisian Masters of Arts with a lever to use in their struggle for improved status. At the same time, the arrival of the Aristotelian corpus, which owed nothing to Christian Revelation and which seemed to challenge certain basic Christian doctrines, aroused misgivings among the theologians. There followed decades of turmoil at Paris, until a synthesis was achieved by the Dominican friar Thomas Aquinas (1225–74), who assimilated Aristotle into Christian teaching so successfully that the 17th century would find it hard to make a separation of the two.

Research was not then the function of a university, and in astronomy the immediate teaching need had been for an elementary textbook that the young students might use. An attempt at this was made in the mid-13th century by John of Holywood (Johannis de Sacrobosco), but his *Sphere* was hopelessly inadequate when faced with the challenge of explaining the motions of the Sun, Moon, and lesser planets. Nevertheless, after the invention of printing the work offered more competent astronomers an excuse to write elaborate commentaries, and in this form it would become one of the best-sellers of all time.

Later in the 13th century an anonymous author made good some of the defects of the *Sphere* with his *Theory of the Planets*. This gave a simple (if only partly satisfactory) account of the Ptolemaic models of the various planets, with clear definitions. Meanwhile, at the court of King Alfonso X of Castile, the old Toledan Tables were replaced by the Alfonsine Tables; modern computer analysis has shown that these tables, which would be standard for the next 300 years, were calculated on Ptolemaic models with only the occasional updating of parameters.

It was not until the 14th century that the Latin West had sufficiently mastered its heritage from the past to be able to break new ground. One development of significance for astronomy came in terrestrial physics, for it was by arguing from the motion of projectiles that Aristotle had most convincingly demonstrated the Earth to be at rest: an arrow fired vertically fell to the ground at the very place from which it had been fired, and this proved that the Earth had not moved while the arrow was in flight.

However, Aristotle was at his least convincing when discussing the physics of projectile motion. An earthly body such as an arrow, he argued, would naturally move downwards towards the centre of the Earth, and its upward (and therefore unnatural) motion must be imposed upon it by an outside force – and not only imposed, but maintained for as long as the arrow was climbing. Aristotle thought the air itself was the only agent available to maintain the upward motion of the arrow; but sceptics had pointed out that this was implausible, since it was possible to fire arrows upwards in the teeth of a gale.

The Parisian masters Jean Buridan (*c*.1295 – *c*.1358) and Nicole Oresme (*c*.1320–82) agreed with Aristotle that a force must be at work, but they rejected any role for the air in this. They argued that an 'incorporeal motive force' must be imposed by the archer on the arrow, a force they termed 'impetus'. Buridan suggested that the heavenly spheres – which though frictionless needed a permanent motive force (angelic intelligences?) if they were to rotate for ever – would spin eternally if endowed with the motive force of impetus at the Creation.

Oresme saw a significant implication of the concept of impetus. *If* the Earth were indeed rotating, the archer as he stood on its surface would be moving with it. As a result, as he prepared to fire the arrow, he would unknowingly confer on the arrow a sideways impetus. Endowed with this impetus, the arrow in flight would travel horizontally as well as vertically, keeping pace with the Earth, and so would fall to ground at the very place from which it had been fired. The flight of arrows, he said, therefore contributed nothing to disputes as to whether the Earth was or was not at rest. Nor, for that matter, did the other arguments traditionally invoked, including those from Scripture. Oresme was of the opinion that the Earth was indeed at rest; but it was no more than an opinion.

The invention of printing in the 15th century had many consequences, none more significant than the stimulus it gave to the mathematical sciences. All scribes, being human, made occasional errors in preparing a copy of a manuscript. These errors would often be transmitted to copies of the copy. But if the works were literary and the later copyists attended to the meaning of the text, they might recognize and correct many of the errors introduced by their predecessors. Such control could rarely be exercised by copyists required to reproduce texts with significant numbers of mathematical symbols. As a result, a formidable challenge faced the medieval student of a mathematical or astronomical treatise, for it was available to him only in a manuscript copy that had inevitably become corrupt in transmission.

After the introduction of printing, all this changed. The author or translator could now check proofs and ensure that the version set in type was faithful to his intentions; and the printer could then multiply perfect copies, to be distributed throughout Europe and available for purchase at prices that were modest compared with the cost of a handwritten manuscript.

Within a few decades the achievements of the Greek astronomers had been mastered and indeed surpassed. The *New Theories of the Planets* of Georg Peurbach (1423–61), the Austrian court astrologer, which appeared in print in 1474, described the Ptolemaic models that underlay the Alfonsine Tables. It also described physically-real representations of these same models, and it may have been the shortcomings of these that led Copernicus to take up astronomy.

In 1460 Peurbach and his young collaborator, Johannes Müller (1436–76) of Königsberg (in Latin, Regiomontanus), met the distinguished Constantinople-born Cardinal Johannes Bessarion (*c.*1395–1472). Bessarion was anxious to see the contents of *Almagest* made more accessible, and he persuaded the two astronomers to undertake the task. Peurbach died the following year, but Regiomontanus completed the assignment. Their *Epitome of the Almagest*, half the length of the original, appeared in print in 1496. It remains one of the best introductions to Ptolemy's masterpiece. The *Almagest* itself was published in an obsolete Latin translation in 1515, in a new translation in 1528, and in the Greek original in 1538. In 1543 a book would appear that surpassed it.

Nicolaus Copernicus (1473–1543) was born in Torún, Poland, and studied at the University of Cracow, where the professors of astronomy had made no secret of their dissatisfaction with the concept of the equant. He then went to Italy, where he studied canon law and medicine, learned some Greek, and developed his interest in astronomy. He is said to have lectured on the subject in Rome around 1500, to a large audience. In 1503 he returned to Poland to take up an administrative canonry of Frombork where his uncle was bishop, and he remained in the diocese for the rest of his life.

Whereas voluminous works by Aristotle were available in Latin in the later Middle Ages, Plato had fared less well: only two minor dialogues had been added to the *Timaeus* that Calcidius had translated in part long ago. All this changed in the Renaissance, as close contacts were resumed with the Greek world, resulting in an influx of Greek scholars into the West prior to the sack of Constantinople in 1453. Plato's dialogues were recovered and admired for their literary qualities, and his mathematical outlook on the cosmos began to supplant that of Aristotle the naturalist. Astronomers began to look for harmony and commensurability in planetary theory, and they failed to find it in the models of Ptolemy, even though the Alfonsine Tables continued to meet the need for tables of reasonable accuracy. In particular, the equant was seen as 'a relation that nature abhors', in the words of Copernicus's disciple Georg Joachim Rheticus (1514–74).

It happened that by now Ptolemy's *Planetary Hypotheses* had been lost to sight, and with it his overall cosmology. The *Almagest* offered models for the individual planets, but Ptolemy's apparent failure to present an integrated picture of the cosmos meant, as Copernicus would put it, that past astronomers

> have not been able to discover or to infer the chief point of all – the structure of the universe and the true symmetry of its parts. But they are just like someone taking from different places hands, feet, head, and the other limbs, no doubt depicted very well but not modelled

from the same body and not matching one another – so that such parts would produce a monster rather than a man.

It was from aesthetic considerations such as these, as much as from specific problems like the absurd variations in the apparent size of the Moon on the Ptolemaic model, that pressure for reform developed – even though in themselves the Ptolemaic models (with improved parameters) did all that could reasonably be asked of them.

There were clues as to the direction a reform might take. Aristotle's custom of citing those whom he intended to refute meant that every student knew of ancient authors who had argued that the Earth was in motion – and Aristotle's rebuttal was no longer so convincing. Peurbach had remarked on how for some unknown reason an annual period cropped up in the model of every single planet. Whatever it was that set Copernicus thinking, not many years after his return from Italy a manuscript by him entitled *Commentariolus*, or *Little Commentary*, began to circulate. In it he outlined his dissatisfaction with existing planetary models, their equants coming in for special criticism. He proposed a Sun-centred alternative, in which the Earth became a planet orbiting the Sun with an annual period, while the Moon lost its planetary status and became a satellite of Earth.

He showed how this led at last to an unambiguous order for the planets (now six in number), in both period and distance. We saw how Ptolemy plausibly assumed that the slower-moving planets were the highest in the sky; but this did not settle the order of Sun, Mercury, and Venus, which accompany each other as they move against the background stars and therefore appear to have the same period of one year. Once this period was accepted as being in fact that of the Earth-based observer, the true periods of Mercury and Venus could be identified, quite different from each other and from that of the Earth; and so an unambiguous sequence of periods could be established.

Copernicus was also able to measure the radii of the planetary orbits as multiples of the Earth–Sun distance; for example, when Venus appears furthest from the Sun (at 'maximum elongation'), the angle Earth–Venus–Sun is a right angle, and by measuring the angle Venus–Earth–Sun the observer can establish the shape of the triangle and therefore the ratios of its sides. The sequence of periods and the sequence of distances proved to be identical. He was later to say of this:

> Therefore in this arrangement we find that the world has a wonderful commensurability, and that there is a sure linking together in harmony of the movement and magnitude of the orbital circles, such as cannot be found in any other way.

This was a powerful consideration in an age dominated by the Platonic search for harmony in the cosmos. Meanwhile, at the more detailed level, Copernicus made a start in *Commentariolus* on developing equant-free models for the planets and Moon.

Years passed, during which Copernicus developed his mathematical astronomy, remote from the intellectual centres of Europe. In 1539, Rheticus, then a teacher of mathematics in the University of Wittenberg, paid him a visit. He found himself enthralled with Copernicus's achievement in developing geometrical models of planetary motions that rivalled those of the *Almagest*, but which were incorporated into a coherent, Sun-centred cosmovision. He secured Copernicus's permission to publish a *First Report* on this work, which he did the following year. He also persuaded Copernicus to allow him to take the completed manuscript (known by its abbreviated Latin title of *De revolutionibus*) to Nuremberg for printing. The task of seeing the work through the press he delegated to a Lutheran clergyman, Andreas Osiander (1498–1552), who with the best of intentions inserted an unauthorized and unsigned preface to the effect that the motion of the Sun was proposed merely as a convenient calculating device; the result was that readers who got no further than this preface had no inkling of the author's true purpose.

The overwhelming bulk of Copernicus's book was concerned with (equant-free) geometrical models of the planetary orbits, and in their daunting complexity these matched the models of the *Almagest*. Proof that they could become the basis of accurate planetary tables – demonstration that the heliocentric approach could pass the practical test – was to come later, with the publication in 1551 of the Prutenic Tables of Erasmus Reinhold (1511–53), which were computed using Copernicus's models. It was in the brief Book I of *De revolutionibus* that Copernicus outlined the striking consequences that follow from the basic assumption that the Earth is an ordinary planet orbiting the Sun.

As we have seen, the list of planets ordered by period was identical with the list of planets ordered by distance. Equally striking, the mysterious 'wanderings' that had given the planets their name now became the obvious and expected consequence of observing one planet from another: Mars seems to go backwards when opposite to the Sun in the sky simply because it is then that the Earth overtakes it 'on the inside'. There is no longer any mystery as to why two planets, Mercury and Venus, are always seen near the Sun, whereas the other three may be observed at midnight; the orbits of Mercury and Venus are inside that of the Earth, while the others are outside.

True, the Earth was anomalous in being the only planet to have a satellite. It was true, too, that the 'fixed' stars did not show the apparent annual motion that one would expect if they were being observed from an Earth in annual orbit (Copernicans retorted that the stars were far away, and their 'annual parallax' was therefore too small for observers to detect). But these were details. The heliocentric universe was a true cosmos:

> In the centre of all resides the Sun. For in this most beautiful temple, who would place this lamp in another or better place than that from

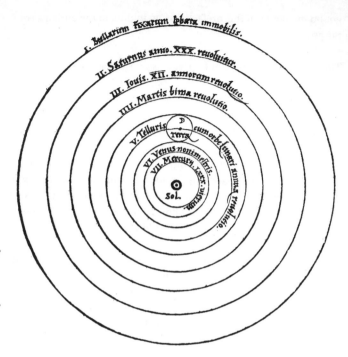

I. Stellarum Fixarum sphæra immobilis.

II. Saturnus anno. XXX. reuoluitur.

III. Iouis. XII. annorum reuolutio.

IIII. Martis bima reuolutio.

V. Telluris cum orbe lunari annua reuolutio.

VI. Venus nonim eliris.

VII. Mercury. reuolutio.

Sol.

11. The outline diagram of the solar system, from Book I of Copernicus's *De revolutionibus*, showing the planets with their approximate periods. Note that no. V, the Earth, is unique in having a satellite, the Moon. Galileo's later telescopic discovery that Jupiter also has satellites helped ease the embarrassment felt by Copernicans because of this anomaly.

which it can illuminate the whole at one and the same time? As a matter of fact, not inappropriately do some call it the lantern of the universe; others, its mind; and others still, its ruler. The Thrice-Great Hermes calls it a 'visible god'; Sophocles's Electra, 'that which gazes upon all things'. And thus the Sun, as if seated on a kingly throne, governs the family of planets that wheel around it.

De revolutionibus is the culmination of the Greek programme to 'save the appearances' of the mysterious planets by geometrical models using combinations of circles rotating with uniform motion. It is an *Almagest* purged of equants, though every bit as complex. It would be a generation before its revolutionary implications sank in.

Chapter 4
Astronomy transformed

Copernicus may have been traditional in his aims and the way he set about achieving them, but his claim that the Earth is in motion had posed a whole range of problems. What moves the Earth? How is it that we have no sensation of movement, and that arrows fired vertically upwards fall to the ground at the place from which they were fired? If we observe the stars as we orbit every six months from one side of the Sun to the other, why can we not detect in them an apparent movement of 'annual parallax'? And how can we explain those passages of Scripture that appear to imply that the Sun is in motion?

Some were misled by the unsigned preface to *De revolutionibus* into believing that Copernicus himself made no claim that the Earth was truly in orbit; merely, that he could 'save the appearances' more successfully by using geometrical models in which the Earth was imagined as moving. Others – including nearly all the competent mathematical astronomers of the next generation – were preoccupied with exploiting these models precisely in order to save the appearances, and neglected the cosmological Book I in which Copernicus's beliefs were clearly stated. Others again looked for some sort of compromise, and among them was a man who was conservative where Copernicus was innovative, and innovative where Copernicus was conservative: Tycho Brahe (1546–1601).

Tycho was born into a noble Danish family, but instead of adopting the pattern of life expected of members of his class in a feudal society, he followed his academic inclinations, among which astronomy figured prominently. In 1563 there occurred a conjunction of Jupiter and Saturn. Because these are the slowest moving of the five planets, their conjunctions, when Jupiter overtakes Saturn, are rare, occurring only once every 20 years, and so are astrologically the most ominous. The teenage Tycho made observations around the time of the conjunction of 1563 and concluded that the date predicted for it in the 13th-century Alfonsine Tables was a month out, while even the modern Prutenic Tables based on Copernicus's models were a couple of days wrong. This, he decided, was unacceptable, and before long he committed himself to the reform of observational astronomy.

Copernicus, like his predecessors, had been content to work with observations handed down from the past, making new ones only when unavoidable and using instruments that left much to be desired. Tycho, whose work marks the watershed between observational astronomy ancient and modern, saw accuracy of observation as the foundation of all good theorizing. He dreamed of having an observatory where he could pursue the research and development of precision instrumentation, and where a skilled team of assistants would test the instruments even as they were compiling a treasury of observations. Exploiting his contacts at the highest level, Tycho persuaded King Frederick II of Denmark to grant him the fiefdom of the island of Hven, and there, between 1576 and 1580, he constructed Uraniborg ('Heavenly Castle'), the first scientific research institution of the modern era.

It had everything: four observing rooms, numerous bedrooms, dining rooms, library, alchemical laboratory, printing press. Elsewhere on the island, there was even a paper mill, so that Tycho was completely self-sufficient in the publication of his work. Four years later, Tycho enlarged his facilities with a satellite observatory,

12. The great mural quadrant of Tycho's observatory at Uraniborg. The wall was aligned north–south, and the observer (barely glimpsed at the extreme right) measured the altitude of a heavenly body as it transited the meridian. One assistant is shown calling out the time of the transit, while another is recording the observation. Tycho, and various features of the observatory, are portrayed in the painting on the wall.

Stjerneborg ('Castle of the Stars'), built partly underground so that its instruments – unlike those in Uraniborg – would be on stable mounts and sheltered from strong winds. But he was loath to admit any limitation in Uraniborg: a second observatory, he said, would prevent collusion between two teams of observers making parallel observations.

Tycho was to remain on Hven until 1597. By then Frederick had been succeeded by Christian IV, who had come to resent Tycho and his arrogant demeanour and who was making life increasingly difficult for him. Tycho therefore quit, and two years later became mathematician to the Emperor Rudolf II in Prague. Tycho had lost his enthusiasm for observations; some instruments remained on Hven and others were kept in store. But he had with him the vast collection of accurate observations of the Sun, Moon, and planets that his teams had made on Hven, and these were to prove decisive for the work of Kepler, who joined him in Prague as assistant and who succeeded him when he died in 1601.

Tycho was the first of the modern observers, and in his catalogue of 777 stars the positions of the brightest are accurate to a minute or so of arc; but he himself was probably most proud of his cosmology, which Galileo was not alone in seeing as a retrograde compromise. Tycho appreciated the advantages of heliocentic planetary models, but he was also conscious of the objections – dynamical, scriptural, astronomical – to the motion of the Earth. In particular, his inability to detect annual parallax even with his superb instrumentation implied that the Copernican excuse, that the stars were too far away for annual parallax to be detected, was now implausible in the extreme. The stars, he calculated, would have to be at least 700 times further away than Saturn for him to have failed for this reason, and such a vast, purposeless empty space between the planets and the stars made no sense.

He therefore looked for a cosmology that would have the geometrical advantages of the heliocentric models but would retain the Earth as the body physically at rest at the centre of the cosmos. The solution seems obvious in hindsight: make the Sun (and Moon) orbit the central Earth, and make the five planets into satellites of the Sun. But the path of discovery was, as so often, tortuous. By 1578 Tycho was thinking of making Venus and Mercury into satellites of the Sun, as had Martianus Capella a millennium before. By 1584 he would have turned all five planets into satellites, except that this would have implied that the sphere carrying Mars intersected with the spheres carrying the Sun.

It was then that he saw the implications of observations he had made back in the 1570s. In November 1572 a star-like object bright enough to be visible in the daytime had appeared in the constellation of Cassiopeia. The heavens (it was thought) had been changeless since the dawn of history, yet the object resembled a bright star. Although he was only 26 and Hven was still in the future, Tycho had already progressed as an observer to the point where he could be certain that the object was celestial rather than atmospheric. Others disputed this, but Tycho was to subject their observations to a critical analysis that eventually settled the issue: the heavens could indeed change.

It might be thought that comets were already ample proof of this; but for Aristotelians, the comets' 'coming to be and passing away' amply demonstrated their terrestrial – or, more exactly, atmospheric – nature. As Aristotle himself had explained, comets resulted from the effect of the rotating heavens on the air and fire that surround the Earth, 'so whenever the circular motion stirs this stuff up in any way, it bursts into flame at the point where it is most inflammable'.

As long as the heavens had been changeless, there had been little reason to dissent from Aristotle's claim that comets were atmospheric. But, in the aftermath of the new star, Tycho harboured doubts. If only Nature would provide him with a bright comet, he would measure its height and establish whether it was atmospheric or celestial. In 1577, when Uraniborg was under construction, Nature obliged; and Tycho established that the comet was moving freely among the planets. As he later realized, this showed that the Earth-centred spheres thought to carry the planets did not exist.

With this, the physical objection to his cosmology disappeared, and in his 1588 book on the comet he presented his system in outline, along with detailed geometrical models for the motions of the Sun and Moon.

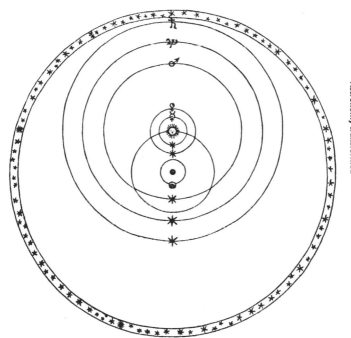

Astronomy transformed

13. **The Tychonic system in outline. The Earth is at the centre, and around it orbit the Moon and the Sun. The Sun is itself orbited by five planets, which it carries around the Earth. The stars are immediately beyond the outermost planet, Saturn. The relative motions are the same as in the Copernican system, and this created great difficulties for Galileo in his campaign in favour of the Copernican theory.**

The stars lay immediately beyond the realm of Saturn, at a distance of some 14,000 Earth radii, so that the Tychonic universe was even more compact than the Ptolemaic.

A number of similar compromises were floated in the decades to come, and many found them attractive. They were to infuriate Galileo Galilei (1564–1642) in his campaign in support of Copernicus because they were so difficult to refute. As professor of mathematics at Padua in the 1590s, Galileo had used the daily and annual motion of the Earth in an attempt to explain the puzzling phenomenon of the tides, but his Copernicanism was less than whole-hearted until the dramatic events of 1609. That summer, when he was in Venice, word came that in Holland instruments with two pieces of curved glass were being used to make distant objects seem near. Curved glass was a traditional source of amusement in fairs because of the distorted image it created, and only when he had reliable confirmation of the rumour did Galileo set about constructing such a device for himself. In August he was able to demonstrate to the Venetian authorities a telescope of 8× magnification, 'to the infinite amazement of all'. Later that year he had improved the magnification to 20×, and it is no coincidence that a few months later he became mathematician and philosopher to the Grand Duke of Tuscany.

Until the invention of the telescope, each generation of astronomers had looked at much the same sky as their predecessors. If they knew more, it was chiefly because they had more books to read, more records to mine. All this now changed. In the coming months and years Galileo saw with his telescope wonders vouchsafed to no one before him: stars that had remained hidden from sight since the Creation, four moons that orbited the planet Jupiter, strange appendages to Saturn that would be recognized as rings only half a century later, moon-like phases of Venus, mountains on the Moon not very different from those on Earth, even spots on the supposedly perfect Sun. He was able to confirm the suggestion of Aristotle that the Milky Way is composed of myriads of tiny stars.

He found that the disc-like shape that stars seem to have when viewed with the naked eye was an optical illusion, so that if Copernicans found themselves forced to banish them to remote regions to avoid the difficulty over annual parallax, they did not thereby have to make them physically huge. Galileo could say of his predecessors, 'If they had seen what we see, they would have judged as we judge'; and ever since his time, the astronomers of each generation have had an automatic advantage over their predecessors, because they possess apparatus that allows them access to objects unseen, unknown, and therefore unstudied in the past.

Galileo lived in an age before scientific journals provided a forum for rapid dissemination of new work, and when leisurely publication in book form was the norm. But his telescopic discoveries could not wait, and he was able to announce the earliest of them within months, in his brief *Starry Messenger* (1610). This was followed in 1613 by *Letters on Sunspots*. Several of his revelations supported Copernicus, none more so than the moon-like sequence of the phases of Venus, which he announced in the *Letters*.

In the Ptolemaic system, Venus was below the Sun in the sequence of planets; furthermore, to 'save' the fact that Venus never appears far from the Sun in the sky, Ptolemy's model required the centre of Venus's epicycle to be on the straight line from the Earth to the Sun. As a result, in the model, Venus always lay somewhere between the Earth and the Sun.

What then if Venus proved to be a dark body illuminated by the Sun? If the Ptolemaic model was correct, the illuminated half would always face partly away from Earth, and so the planet would never appear to us as a circle of light, like the Moon when full. For Copernicus, by contrast, Venus orbited the Sun on a path inside that of the Earth. When near to Earth it would appear to have a crescent shape because its illuminated half would be facing away from Earth, as in the Ptolemaic model; but when away on the far side of the Sun it would appear full.

And this was just what Galileo had witnessed. Ptolemy was wrong, disproved by a decisive observational test. Copernicus was therefore right – or so Galileo would have us believe. But the phases of Venus tell us only about the *relative* positions of Sun, planet, and Earth, and nothing about which of the three is physically at rest; and the relative motions are broadly the same in both the Copernican and Tychonic systems. The Tychonic system was therefore unscathed by Galileo's discovery.

To Galileo this was most unwelcome, because Ptolemy had long since been discarded by those who believed the Earth to be at rest in favour of the Tychonic, or one of the 'semi-Tychonic', systems. His telescopic discoveries had made him an ardent propagandist for Copernicus; but proving Ptolemy wrong was easier than proving Copernicus right. And so he continued to act as though the choice was still between Ptolemy and Copernicus: as late as 1632 he gave his Copernican manifesto the title of *Dialogue on the Two Great World Systems, the Ptolemaic and Copernican*.

To resolve the physical objections – how is it that we Earthlings are hurtling through space, deluded all the while into believing we are in fact on *terra firma*? – Galileo created a new conception of motion. Motion – change of all kinds, of which change of place is only one – had been fundamental to Aristotelian philosophy, for a natural body expressed its nature by how it behaved, how it moved. In Aristotle's view, motion demanded explanation, rest did not.

Galileo, to the contrary, set out a new way of looking at the world, in which it was change of motion – acceleration – that called for explanation, while steady motion (of which rest was now merely a special case) was a state that needed no explanation. He imagined a ball rolling on the surface of a perfectly smooth, spherical Earth, and saw no reason why the ball should ever come to rest: it would remain indefinitely in a state of uniform motion, rolling around the centre of the Earth. Similarly, the Earth itself orbited the centre of the solar system, in a state of uniform motion, which was why Earthlings were unaware of their movement.

Galileo had a gift for friendship but also a gift for enmity, and the view that the Earth moves had long been seen as in apparent contradiction to certain Scriptural phrases. In 1613 Galileo wrote a semi-public letter to a friend that was to become the basis of his *Letter to the Grand Duchess Christina* (written in 1615 but not published until 1636). This is now recognized as a classic statement of the traditional Catholic position – that the Bible teaches us how to go to heaven, not how the heavens go – but the period of the Counter-Reformation was no time for a layman to pronounce on the interpretation of Scripture. In 1614 a preacher mounted an attack on him by turning a text from the Acts of the Apostles into perhaps the best pun in ecclesiastical history: 'Ye men of Galilee, why stand you gazing up to heaven?' Galileo, ignoring warnings from friends, stood his ground; the dispute escalated, and the Vatican became involved. Eventually, in February 1616, Galileo was interviewed by the saintly Cardinal Robert Bellarmine. Bellarmine, who was prepared to revise his traditional position on the stability of the Earth but only when compelling proof was forthcoming, notified Galileo privately that he was no longer permitted to believe that the Copernican system was true, or to defend this belief.

Time passed, and in 1623 the election of a new Pope who had been Galileo's friend and supporter encouraged him to resume his campaign in support of Copernicanism, and this led eventually to the publication of his *Dialogue* in 1632. Exactly what it was that so upset the Roman authorities is still debated, but the outcome was that Galileo was condemned to house-arrest. Comfortable though this in fact proved to be, his condemnation was a setback to astronomy in Catholic countries, and the many Jesuit astronomers found themselves expected to support the Tychonic system or a similar compromise.

There was in Galileo's makeup a certain laziness, in particular a reluctance to engage in hard mathematics, and this cost him dear in his campaign as Copernican propagandist; for he remained oblivious to the service rendered to the Copernican cause by a contemporary who saw the planets as bodies driven round by forces emanating from the massive, central Sun – and who was thereby transforming astronomy from applied geometry to a branch of physics, from kinematics to dynamics. Johannes Kepler (1571–1630) was born at Weil der Stadt near Stuttgart and studied at the University of Tübingen, where his fair-minded teacher in astronomy set out the pros and cons of the various cosmologies on offer, including that of Copernicus. Kepler then commenced studies in theology, but in 1593 the authorities nominated him to teach mathematics in Graz, and with reluctance he complied.

Settled in Graz, he began to puzzle over the structure of the universe, which he saw as created by God, the great geometer. Copernicus, he believed, had discovered the basic layout of the universe, but not the reasons that had motivated God in His selection of this particular universe from among the possible options. In particular, why had God created just six planets, and why had He given the five spaces between them the sizes that they have? Eventually it occurred to Kepler that the number of spaces equalled the number of the regular solids (pyramid, cube, and so on), shapes that must have a profound appeal for any geometer, human or divine. And so Kepler investigated the geometry of nests of six concentric spheres, each pair separated by one of the five regular solids in such a way that the inner sphere of the pair touched the planes of the solid while the outer sphere passed through its vertices; and eventually he identified a particular nest that reproduced, to a reasonable approximation, the radii as calculated by Copernicus.

This did not address the question of the speeds of the planets, and Kepler took the epoch-making step of approaching this problem in

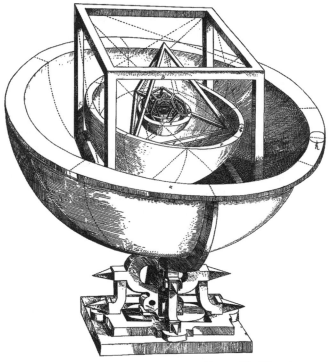

14. The geometrical relationships embodied by God in His universe, according to Kepler's *Cosmographic Mystery* (1596)

terms of the physical influence of the central – and massive – Sun. After all, Copernicus had shown that the further a planet was from the Sun, the slower it moved in its orbit. Perhaps this was because the Sun was the cause of the planets' movement, and the effectiveness of this cause diminished with increasing distance.

It would be more than two decades before Kepler would establish the actual pattern of the planetary speeds, but his early speculations, published in *Cosmographic Mystery* (1596), served notice on the astronomical community that a new talent had appeared on the scene. Kepler sent a copy to Galileo, urging him to come out in support of Copernicus, but he received only a polite response. Tycho, however, invited Kepler to join him on Hven, notwithstanding the fact that Kepler's book was the first that was irredeemably heliocentric. Kepler decided against moving to this remote island; but early in 1600, when Tycho had relocated to Prague, Kepler decided to pay him an exploratory visit, and there for three months he worked on the orbit of Mars. Except for Mercury, which is hard to track because it spends so much time lost in the glare of the Sun, Mars has the orbit that departs most from the circular and therefore is the most difficult to 'save' with a geometry of circles. After the three months were up, Kepler returned to Graz, but he was soon back with Tycho, who had only a short time to live. Within a year Kepler found himself Tycho's successor.

Kepler's 'warfare' with Mars, the god of war, lasted for several years. His campaign was, he said, based on the Sun-centred vision of Copernicus, the observations of Tycho Brahe, and the magnetical philosophy of William Gilbert (1544–1603), whose *On the Magnet* (1600) argued that the Earth itself was a vast magnet.

Kepler is that rarity, an honest scientist who in his publications does not launder his account of his research to make the path to his conclusions seem direct and untroubled. Kepler requires his reader to follow him in the maze of calculations, and he is no fit subject for any sort of Introduction, let alone a Very Short one. But the essential point is clear: Kepler abandoned the geometrical models of traditional astronomy, the study of *how* planets move, in favour of physics, the study of what forces cause them to move as they do.

This shift from kinematics to dynamics had been almost inevitable, once Tycho had demonstrated the non-existence of the heavenly spheres commonly thought to carry the planets round. Why these spheres continued to spin had been of minor interest – most held that angelic intelligences drove them on, Buridan had postulated an impetus given to each sphere at the Creation, while Copernicus thought that all natural spheres naturally spun. But take away the planetary spheres, and you are left with planets that are isolated bodies in orbit. What could it be that drives them round, almost in the manner of projectiles? Once this question was given centre-stage, the success of the heliocentric hypothesis was assured: it made dynamic sense for the relatively small Earth to orbit the massive Sun, but not vice versa.

In 1609 Kepler published his solution to the problem of Mars, in a work whose challenging title proclaimed the reorientation of astronomy: *New Astronomy Based upon Causes, or Celestial Physics, Treated by Means of Commentaries on the Motions of Star Mars*. Early on, Kepler had come to understand that he must take the real, physical Sun as the hub of his solar system, and not some point that was geometrically convenient. Equally, he must give a combined account of the motions of the planet in longitude and latitude; it was no longer acceptable to devise two geometrical models, each doing the job for one or other coordinate, but incompatible with each other.

Tycho's data were numerous enough to provide generous amounts of information about Sun, Earth, and Mars when they were in special configurations. Thus when Mars was exactly opposite to the Sun in the sky, an Earth-based observation could do duty for an observation from the Sun at the centre of the system; and Tycho had records of a number of these. They were accurate: when Kepler arrived at a model with a circular orbit that 'saved the appearances' to as little as 8 minutes of arc (good enough to match the observations of any earlier observer), Kepler knew that because Tycho's accuracy was much better than 8 minutes, the model was not satisfactory and he must reject it. Yet Tycho's observations were not (so to say) *too* accurate: the orbit of the real Mars is disturbed by the pulls of other planets, and so hypothetical observations of perfect accuracy would have prevented Kepler from reaching his eponymous laws.

Gilbert had argued that the Earth is a large magnet; perhaps the Sun was an even greater one. As the planets all orbit the Sun in the same direction, and move less urgently the further they are from the Sun, Kepler was led to think of the Sun as a rotating body that sends out into space a magnetic influence that pushes the planets round, the influence naturally being most effective on the nearest planets. Without this continual influence, Kepler believed, each planet would stop dead in its tracks – inertia in our everyday sense of the word.

But there was more to it than this, for the planetary orbits are not simple circles. The planets vary their distances from the Sun, and to account for this Kepler introduced into the Sun a second force that attracted the planet for part of its orbit and repelled it over the rest.

These physical intuitions guided Kepler in his analysis of Tycho's observations. As it happened, Kepler discovered his 'second law', which tells us about the speed of the planet in its orbit, before the 'first law', which tells us what the orbit is. According to the first law, the planet moves in an ellipse with the Sun at one focus. Ellipses had been well understood by geometers since the 2nd century BC, and it was a masterstroke to define a planetary orbit, not with a Ptolemaic-type model bristling with epicycles, eccentres, and equants, but through a single, very familiar curve. But in arriving at the law Kepler's physical intuition had at one stage proved something of a barrier – for an ellipse has symmetry about the minor axis, whereas Kepler's first law tells us that the orbit is highly unsymmetric about this axis, the Sun being at one focus and the other being 'empty'.

We know the second law in the variant form that Newton later found to be a *consequence* of his law of gravitational attraction: a line from the Sun to the planet traces out equal areas in equal times. Mathematicians were barely able to deal with such a bizarre formulation, and they preferred other variants that were mathematically tractable, and could hardly be distinguished observationally from the area law: that the planet moves with a speed inversely proportional to the Sun–planet distance; or that the planet moves with apparently uniform speed when viewed from the empty focus. In our earlier discussion of the Ptolemaic equant in a circle, we saw why its equivalent in an ellipse – uniform speed viewed from the empty focus – gives such a good approximation to the area law.

New Astronomy set out how Mars – and by implication the other planets – moved, and what caused it to do so. But what of the overall pattern of the planetary system, which Kepler had first tackled in *Cosmographic Mystery*? This was one of many themes of *The Harmony of the World* (1619), along with such topics as the music generated by the planets in their orbits. Copernicus had rejoiced to find that the further a planet was from the Sun, the longer it took to complete a circuit; Kepler could now announce the formula that ensured this: the square of the period of a planet is in a fixed ratio to the cube of the radius of its orbit.

Kepler was by now engaged in making his work accessible to a wider readership, setting it out in a user-friendly question and answer form. The title of his *Epitome of Copernican Astronomy*, which appeared piecemeal between 1618 and 1621, paid tribute to his ultimate source of inspiration; but Copernicus would have been baffled to discover that his geometrical astronomy had now migrated to become a branch of physics.

Planetary theories had always been subjected to the ultimate practical test: could they be used to generate accurate tables? Tycho's own interest in astronomy had been kindled by shortcomings in the Copernican Prutenic Tables, and when Tycho had first presented the young Kepler to the Emperor Rudolf, the Emperor commissioned Kepler to work alongside Tycho on tables on which astronomers might at last rely. In 1627, when Rudolf and Tycho were long dead, Kepler published the Rudolphine Tables. Kepler too was dead when the French astronomer Pierre Gassendi became the first observer in history to see a transit of Mercury across the face of the Sun. The forecast in the Rudolphine Tables was 30 times more accurate than that in the Prutenic: Kepler's elliptical astronomy had passed the test. Yet his physical intuition, of planets that would instantly come to halt without the incessant urging of the Sun, was wholly implausible, and the formulation of the second law – the crucial one that defines the varying velocity of the planet in its orbit – was confused and confusing. Kepler had replaced numerous circles with single ellipses, and he had encouraged astronomers to see their discipline as 'celestial physics'. But the true dynamics of the planetary system was as mysterious as ever.

Chapter 5
Astronomy in the age of Newton

The outlook of the later Middle Ages had been dominated by Aristotle, and that of the Renaissance by Plato. But in the period following, the 'mechanical philosophy', alternatively known as the 'corpuscular philosophy', became increasingly attractive. It had originated with the Greek atomists, who explained the different qualities we perceive in the bodies around us as being the ways our senses interpret the movements of unchanging particles; and this appealed to an age that found a refreshing clarity about explanations that used concepts like speeds and shapes, concepts that were mathematical at least in principle. Machinery was becoming ever more ingenious – witness the great clock in the cathedral in Strasbourg. But in these machines, complex effects were being produced by simple (and intelligible) means: matter – cogs, weights, and so forth – in motion.

God was now the great clockmaker, His creation hugely complex in structure, but intelligible simply as matter in motion. Galileo was one of many attracted by this revival of ancient ideas, but it was his younger contemporary, René Descartes (1596–1650), who carried the mechanical philosophy to its extreme. At the school at La Flèche his Jesuit teachers introduced him to Galileo's telescopic discoveries within months of their announcement; more importantly, they instilled in him a profound admiration for the certainty to be found in geometrical theorems. For Descartes, there was a vast gulf between the certainly true and what was only very probably true; and he decided that to bridge the gulf one must imitate the reasoning of the geometers. Furthermore, it was the geometers who had always had the correct understanding of space: the infinite, homogenous, undifferentiated space of Euclid was not the idealized abstraction it had been taken to be, but the space of the real world.

As a philosopher, Descartes was ruthless. Unlike Galileo, who could never mention Aristotle without getting involved in a squabble, Descartes dismissed Aristotle with contempt and set about creating his own philosophy. Until now, all discussion had referred back to Aristotle, favourably or otherwise; soon all discussion would be referred back to Descartes.

In the Cartesian universe there were no longer any privileged places, such as the centre of the Earth or the centre of the solar system, that were different from ordinary places because it was around them that motions took place. Analysing the fundamental concept of matter *per se*, Descartes rejected properties like colour or taste that belong to some matter but not to all, and ended by concluding that matter and space were in many respects identical. This being so, space without matter – a vacuum – was an impossibility: the world is a plenum. Furthermore, since space was uniform, so was matter. This meant that the differences we perceive between this material object and that are due entirely to how the (uniform) matter is moving in the two spaces involved: motions are everything when it comes to understanding the universe.

Because God lives in the eternal present, He conserves this particular matter in the motion that it has at this instant of time, travelling in a specific direction with a given speed: the law of rectilinear inertia. And just as He conserves the total space of the universe, so He conserves the total amount of motion in the universe. This allows us to establish the laws that govern the transfer of motion from one object to another.

Because the universe is packed with matter, rectilinear inertia is a tendency rather than a reality. In practice, matter can move only if the matter ahead of it (and the matter behind it) also moves. As a result, matter normally moves in vortices, or whirlpools; these act rather like a centrifuge, with some of the matter forcing its way towards the exterior, and other matter as a result being pushed towards the centre. The latter we see as self-luminous, and we perceive the great assemblages of this luminous matter as the Sun and stars. The Sun is therefore nothing more than our nearest star, and similar stars are scattered everywhere throughout the infinite universe.

The Sun is at the centre of a great solar vortex that carries the planets round; these planets tend to fly off at a tangent, but they are constrained into closed orbits by the rest of the matter in the vortex. Among the planets is the Earth, itself the centre of a lesser vortex that carries the Moon. The Moon is therefore transported in both the solar and the terrestrial vortices, and because of this Newton was to find its motion hard to calculate. But although Descartes was a mathematician, and wrote in a letter that 'My physics is nothing but geometry', *The Principles of Philosophy* (1644), in which he set out his ideas, consists only of words and has no mathematical equations. Words are vague and malleable, and *The Principles* proved so adaptable that it could explain almost everything, while predicting almost nothing. The book could be understood by the innumerate; and the world-picture it expounded was to prove immensely attractive to devotees of the Parisian *salons*.

In Oxford and Cambridge in the later decades of the century, Aristotle still officially held sway, though within the individual colleges enterprising tutors were able to introduce their charges to the novelties of Cartesian philosophy. London, however, was home to the Royal Society, founded in 1660, and the Fellows were heirs to the 'magnetical philosophy' of William Gilbert, who had so greatly influenced Kepler's thinking. A leading figure in the Society was John Wilkins (1614–72), who in 1640 had published a second edition of his *Discovery of a World in the Moone*, in which he argued that Moon travel was theoretically possible because the magnetic influence of the Earth diminished with height: 'it is probable, that this magneticall vigor dos remit of its degrees proportionally to its distance from the earth, which is the cause of it.'

Early in the 1660s, Robert Hooke (1635–1703), Curator of Experiments to the Royal Society, even carried out tests to see whether the pull of the Earth was less at the top of a cathedral than it was at ground level. The results were of course inconclusive; but in the years that followed, Hooke steadily generalized his thinking about magnetism as an explanation for what we see about us in the solar system. By 1674 he was able to set out the stage he had reached in three remarkable 'Suppositions':

> First, That all Coelestial Bodies whatsoever, have an attraction or gravitating power towards their own Centers, whereby they attract not only their own parts, and keep them from flying from them, as we may observe the Earth to do, but that they do also attract all the other Coelestial Bodies that are within the sphere of their activity . . . .

Hooke believed that all the bodies of the solar system attracted other bodies – more exactly, those 'within the sphere of their activity' – by a force identical to the gravity that held the parts of the Earth together. About rectilinear inertia he was admirably clear:

> The second supposition is this, That all bodies whatsoever that are put into a direct and simple motion, will so continue to move forward in a streight line, till they are by some other effectual powers deflected and bent into a Motion, describing a Circle, Ellipsis, or some other more compounded Curve Line.

> The third supposition is, That these attractive powers are so much

the more powerful in operating, by how much the nearer the body wrought upon is to their own Centers.

But did they vary inversely with the distance itself ($f \propto 1/r$), or with the distance squared ($f \propto 1/r^2$), or what? Hooke could not say; he regarded the answer as relatively unimportant, merely one of the loose ends to be left to mathematicians.

The inverse-square law was an obvious candidate, because the brightness of a heavenly body falls off with the square of the distance. But there was a more cogent reason. Analysis of the dynamics of circular motion – of a stone whirled around in a sling, for example – combined with Kepler's third law of planetary motion suggested that if the planets were all to move around the Sun in strictly circular orbits and at steady speeds, the entire pattern of these orbits could be explained as resulting from solar attraction that diminished with the square of the distance. Yet the true orbits of the planets were elliptical; could it be shown that elliptical orbits likewise result from an inverse-square law of attraction?

By 1684 opinion in London had hardened: the inverse-square law was the answer, but no one could handle the mathematics needed to demonstrate this. Might Isaac Newton (1642–1727), the highly gifted but secretive mathematics professor in Cambridge, be able to do so? Edmond Halley (c.1656–1742) plucked up his courage and bearded Newton in his den. What, he asked, would be the shape of the orbit of a planet attracted to the Sun under an inverse-square law? Newton unhesitatingly gave the reply Halley was hoping for: an ellipse.

Newton had entered Cambridge in 1661, and by the middle of the decade he too had successfully analysed the dynamics of strictly circular motion. But he was facing serious problems in trying to understand the planetary orbits. In the Cartesian universe (which Newton then accepted as a true representation of reality), the Moon was carried in both the solar and the terrestrial vortices, and this made mathematical analysis difficult. As to the planets, there were several variants of Kepler's second law in circulation, observationally almost identical but conceptually worlds apart. Newton himself tried working with a number of equant versions of the law. Eventually, however, he came across a book that set out the law in the area formulation that we know today.

In 1679 Newton was still struggling to make sense of vortices, and still confused in his analysis of the dynamics of orbital motion, when he received a letter from Hooke. Hooke was now Secretary of the Royal Society, and eager to involve the Cambridge mathematician in its activities. He invited Newton to consider the consequences of 'compounding the celestiall motions of the planets of a direct motion by the tangent [inertial motion] and an attractive motion towards the centrall body'. Hooke saw orbital motion, not as the outcome of a struggle between centrifugal and centripetal forces, but as the effect of an attractive force on motion that would otherwise continue in a straight line.

Hooke also told Newton about his attempt (see Chapter 6) to prove the motion of the Earth by measurement of annual parallax. In his reply, Newton offered a new approach to the traditional 'proof' that the Earth must be at rest, because arrows fired vertically fell to the ground at the place from which they had been launched – or, equivalently, a stone dropped from a tower fell to the ground at the base of the tower. Newton pointed out that the top of the tower was further than the base from the centre of the Earth; and that since the Earth was in fact spinning, the stone when still at the top of the tower was travelling horizontally more rapidly than the ground at its base. Therefore, he argued, since the stone would retain its horizontal velocity while falling, it would in fact strike the ground ahead of the tower. And he went on to discuss how the stone would continue to move, in the imaginary situation in which it had the power to pass through the Earth unimpeded. By doing this Newton converted a problem of free fall into one of orbital motion.

Hooke had the satisfaction of pointing out a mistake in Newton's analysis, though an imagined passage through the Earth was too fanciful for his taste. However, he did reveal to Newton his commitment to an inverse-square law: 'my supposition is that the Attraction always is in a duplicate proportion to the Distance from the Center Reciprocall.'

Newton's reaction to any hint of criticism was to withdraw into his shell, at the same time secretly devoting his energies to establishing the truth of the matter. Although Hooke's proposed scenario was, he thought, divorced from the real (Cartesian) universe which was crammed with matter, Newton pursued the mathematical analysis – and made the extraordinary discovery that Hooke's planets would move in ellipses with the Sun at a focus, and that the line from the Sun to the planet would trace out equal areas in equal times, exactly as Kepler had declared the real planets to do. Could it be that Hooke's was the real world – a world largely empty, in which isolated bodies somehow influenced each other across the intervening space by attraction – and Descartes's plenum the imaginary?

We know little of how Newton's thinking developed between 1679 and the visit from Halley in 1684, except for a confused exchange of letters between Newton and the Astronomer Royal, John Flamsteed (1646–1719), as to whether a comet seen approaching the Sun in November 1680 was identical with the comet seen leaving the Sun the following month (it was); and if so, what had happened to it meanwhile and why. Newton at one point suggested that it might have been a single comet that 'fetched a compass round the Sun' – that it had gone around the back of the Sun – and he may already have been thinking that the comet's path was the result of the attraction of the Sun; but we cannot be sure. Certainly the visit from the suitably deferential and tactful Halley encouraged Newton to promise him written proof that elliptical orbits would result from an inverse-square force of attraction residing in the Sun. The drafts grew and grew, and eventually resulted in *The Mathematical*

Principles of Natural Philosophy (1687), better known in its abbreviated Latin title of the *Principia*. Contemporaries recognized in the title a challenge and rebuke to the wordy and fanciful *Principles of Philosophy* of Descartes.

The first draft ran to a mere nine pages. It analysed the orbit of a body moving with inertial motion in empty space, under the influence of a pull to a 'centre'. Such a body would obey Kepler's area law. If the pull was inverse-square, then the orbit would be a conic section – an ellipse, parabola, or hyperbola. If bodies moved in elliptical orbits with the pull to the focus, then the orbits would obey Kepler's third law; and vice versa. All three of Kepler's laws (the second in 'area' form), which had been derived by their author from observations, with the help of a highly dubious dynamics, were now shown to be consequences of rectilinear motion under an inverse-square force.

Newton had not yet arrived at the point of seeing attraction as a mutual force between celestial bodies large and small, as Hooke had done much earlier; and this is curious, since his draft stated that Kepler's third law applied to the Galilean moons of Jupiter, and to the five moons of Saturn (Titan had been discovered by Christiaan Huygens in 1655, and Gian Domenico Cassini had since discovered four more). These moons were therefore attracted by their parent planet, and one wonders why, if Saturn pulled Titan, it did not also pull the Sun. Perhaps the same thought occurred to Newton, for in his next draft attraction was universal.

In Newton's mind, the Cartesian universe, crammed full of matter, in which bodies constantly impacted on each other, had now given place to one that was almost empty, and in which bodies moved with rectilinear inertia modified by the attractions of all the other bodies – attractions that somehow reached across empty space. Newton was justifiably aghast at the complexity of the mathematical challenges that resulted, not least in the study of the ways in which the Moon behaves under the competing pulls of

Earth and Sun. For their part, Continental mathematicians would be shocked at Newton's retrograde step in invoking a mysterious 'attraction', for which he offered no mechanism and which sounded for all the world like the reintroduction of the dubious 'sympathies' and other 'occult qualities' that the mechanical philosophy had only recently eradicated.

After two millennia of observation and analysis, the how and the why of planetary orbits were at last understood, even if the nature of the inverse-square force remained mysterious. But what of comets? Newton was now certain that the two comets seen late in 1680 were one and the same, and that it had indeed 'fetched a compass' around the Sun. Comets, he concluded, were part of the same universal pattern, and in the *Principia* he showed that their orbits were conic sections (though not necessarily ellipses) and that they too obeyed Kepler's area law. This opened up the possibility that a comet moving in an elongated ellipse would regularly return to the solar system.

Hooke had long since suspected that the Earth's pull on a falling stone was the same as its pull on celestial bodies, and the same thought now occurred to Newton. But to compare the pull of the Earth on a stone with its pull on the Moon, he faced a mathematical challenge. He would have to combine the pulls on the stone of all the bodies that make up the Earth, pulls operating over distances that ranged from a few feet to thousands of miles. Leaving aside the question of how a force could operate with equal effectiveness through thin air and through miles of rock and earth – some of his followers would admit this could happen only by the direct fiat of the Creator – Newton proved the remarkable theorem that the combined pulls equalled the pull of the entire Earth imagined as concentrated at its centre.

He could now compare the total pull of the Earth on the stone (effectively over a distance of one Earth radius) with the pull by which the Earth drew the Moon into a closed orbit (at a distance of

60 Earth radii). He found the ratio was indeed about 60^2:1. Earth and sky obeyed the inverse-square law of attraction.

As the drafts of *Principia* multiplied, so too did the number of phenomena that at last found their explanation. The tides resulted from the difference between the effects on the land and on the seas of the attraction of Sun and Moon. The spinning Earth bulged at the equator and was flattened at the poles, and so was not strictly spherical; as a result, the attraction of Sun and Moon caused the Earth's axis to wobble and so generated the precession of the equinoxes first noticed by Hipparchus. Several 'inequalities', or irregularities, in the motion of the Moon had been detected – one by Ptolemy and others by Tycho – and these too Newton was able to explain, qualitatively if not quantitatively.

Our satellite is easy to observe, and the pulls on it are simple to state, but highly complex to analyse mathematically. Newton set the agenda for 18th-century mathematicians of the highest talent: to show that the observed lunar motions can be fully explained by the inverse-square law. The historical investigation of these increasingly successful attempts is not for the faint-hearted; fortunately for us it belongs to the history of applied mathematics rather than to the history of astronomy.

Newton was able to use the observed motions of the moons of Earth, Jupiter, and Saturn to calculate the masses of the parent planets, and he found that Jupiter and Saturn were huge compared to Earth – and, in all probability, to Mercury, Venus, and Mars. It seemed, therefore, that the two massive planets had been located at the outer reaches of the solar system, where their powerful gravitational attraction would do least harm to the stability of the solar system. In time, however, even this providential arrangement would begin to experience perturbations, and the system would then 'want a reformation': Providence would step in to restore the original order, and thereby demonstrate God's continuing care for mankind.

Some Continentals, notably Gottfried Wilhelm Leibniz (1646–1716), who agreed with Newton that God was the great clockmaker and saw the universe as a masterpiece of machinery, were scandalized that Newton thought God so poor a workman as to need to rectify his blunders by working miracles in this way. But for Newton this was all part of God's plan from the beginning; He had entered into a servicing contract with the universe, to demonstrate His continuing care for His Creation.

Other Continentals found the concept of attraction retrograde: true, Newton had used this supposed force to explain many movements, but the force itself was so far inexplicable. Could it perhaps be explained on Cartesian principles? The antiquated geometrical formulation of the theorems of Newton's *Principia* had been enough to deter all but a few would-be readers. It was only when popularizations began to appear in the early decades of the 18th century, and more especially when Continental mathematicians successfully exploited the Newtonian programme and explained more and more aspects of the complex behaviour of the Moon, that the merits of attraction become indisputable. Any remaining doubts were dispelled in 1759 by the reappearance of a comet.

According to Cartesian physics a comet was a dead star whose own vortex had collapsed, and which then wandered from one vortex to another, although if it penetrated far enough into a vortex it might remain there as a planet. Newton, however, had claimed that comets obeyed Kepler's laws (in their generalized form), and that a comet whose orbit was an elongated ellipse would regularly reappear. Halley therefore searched the historical records, looking for three or more comets all with similar orbital characteristics, whose appearances had been separated in time by the same number of years or multiples thereof; and he found that the comets of 1531, 1607, and 1682 seemed to fit the bill. In 1695, he told Newton that he thought these were reappearances of the same comet.

However, the intervals separating them, though similar, were not identical. Halley realized that this was because the orbit would have been modified, whenever the comet passed near a major planet during its passage through the solar system and in so doing experienced the planet's gravitational pull; and he predicted the same comet's return 'about the end of the year 1758 or the beginning of the next'.

Could it be that these portentous apparitions were, after all, as lawlike as planets? In the summer of 1757, Alexis-Claude Clairaut (1713–65) and two associates laboured against the clock to calculate in greater detail how the orbit of the comet would have been modified in 1682 as it passed close to Jupiter when departing the solar system; and finally they were able to predict that the returning comet would swing around the Sun within a very few weeks of mid-April 1759.

A newly arriving comet was indeed seen on Christmas Day of 1758, and it rounded the Sun on 13 March 1759. Crucially, the characteristics of its orbit were closely similar to those of the three comets Halley had studied: all four comets were one and the same. To the astonishment of astronomers and public alike, Newtonian mechanics had predicted the return of 'Halley's Comet' after an interval of three-quarters of a century.

Meanwhile, much mathematical effort was being expended on analysing the complex behaviour of the Moon. This was motivated in part by mathematical curiosity, but there was also a much more serious purpose. The lives of sailors at sea depended on their knowing where they were, especially at night. To determine the ship's latitude was relatively straightforward: the navigator measured the altitude of the north celestial pole at night (or, less directly, the altitude of the Sun at midday). Longitude – the time-change all too familiar to air travellers today – was much more difficult, for how was one to compare local time with a standard time (today, Greenwich Mean Time)? By the early 18th century, pendulum clocks were performing acceptably well on land, but they were useless at sea.

From time to time down the centuries, attempts had been made to exploit a suggestion of Hipparchus, that the difference in longitude between cities could be determined by comparing the local times of an eclipse of the Moon, viewed simultaneously from the two locations; but such eclipses were too rare to be of use to navigators. Galileo had proposed instead the eclipses of Jupiter's moons, which are much more common; and later in the 17th century accurate tables of the Jovian satellites allowed this method to be used successfully on land. But such eclipses – still rare enough in all conscience – were well-nigh impossible to observe from on board ship.

Other methods were tried that varied from the near-hopeless to the downright bizarre, and eventually the serious alternatives reduced to two: the development of a chronometer that would keep accurate time at sea, and the use of the Moon's rapid movement against the background stars as the analogue of the movement of a clock's hour-hand against the hour-numerals of the dial. The prize offered by the British Parliament for a practical solution to the problem of longitude at sea would make the recipient rich beyond the dreams of avarice.

The chronometer was work for clockmakers, men who worked with their hands, chief among them John Harrison (1693–1776). Meanwhile, university-trained astronomers and mathematicians struggled to perfect the method of 'lunar distances'. To implement this method, the navigator would first have to determine the Moon's current position in the sky – in practice, its position relative to nearby stars. For this he needed an accurate star catalogue, and an accurate instrument with which to measure the angles between the Moon and convenient stars. He then required reliable lunar tables that would convert this observed position of the Moon into standard time, which he could compare with his local time to give him his longitude. Errors in star positions, in the measurement of the angles, and in the lunar tables would all increase the distance between where the ship in fact was and where the navigator reckoned it to be, and so it was crucial that each of the three be reduced as far as was humanly possible.

The Royal Observatory at Greenwich was founded in 1675 expressly to meet navigators' need for an accurate star catalogue; and the posthumous publication in 1725 of Flamsteed's 'British Catalogue' of 3,000 stars, which improved on Tycho's naked-eye star catalogue by a whole order of magnitude, was the first Astronomer Royal's fulfilment of this need. The requirement for an accurate instrument for measuring angles, suitable for use at sea, was solved by the invention in 1731 of a double-reflection quadrant, ancestor to the sextant. It was now up to mathematicians – all of them, as it happened, French or German – to perfect Newton's lunar theory so that accurate tables of the Moon's position could be calculated months in advance and supplied to navigators. Eventually the Göttingen professor Tobias Mayer (1723–62) developed tables good enough to earn his widow £3,000 of the prize on offer in Britain, and these allowed the then Astronomer Royal, Nevil Maskelyne (1732–1811), to publish in 1766 the first of the annual volumes of *The Nautical Almanac*.

Meanwhile, however, Harrison was producing a succession of masterly chronometers, the first of which was taken for trial to Lisbon and back in 1736. The results were encouraging and Harrison was awarded £250 to fund further research and development. Things continued in this way for nearly 30 years, until in 1764 Harrison sailed with his fourth chronometer to Barbados and back, after which he was awarded half of the £20,000 originally offered as prize money. As soon as suitable chronometers could be constructed in quantity, they became the preferred solution to the problem of longitude, and astronomers found themselves with a new role, manning observatories at major ports and dropping a time-ball at noon (or 1 p.m.) so that navigators could check their chronometers before setting sail. Harrison's own chronometers – poetry in motion – may be seen in operation today in the National Maritime Museum at Greenwich.

Newton had seen the curiously large gap separating the small inner planets from the massive Jupiter and Saturn as evidence of Providence's concern to preserve the solar system from disruption, but Kepler had earlier toyed with the idea that the gap was occupied by a planet as yet undiscovered. By the 18th century, references to 'the known planets' (with the implication that there might be others still unknown) were not uncommon, and speculations about a possible 'missing' planet were fuelled by the discovery of a curious arithmetical pattern in the distances of the planets from the Sun. In his *Elementa astronomiae* of 1702, the Oxford professor David Gregory (1659–1708) had put these distances as proportional to 4, 7, 10, 15, 52, and 95; and by slightly modifying two of the numbers, Johann Daniel Titius (1729–96) of Wittenberg made them equal to 4, 4 + 3, 4 + 6, 4 + 12, 4 + 48, and 4 + 96. These numbers have the form $(4 + 3 \times 2^n)$. The proposed pattern was enthusiastically adopted by the young German astronomer Johann Elert Bode (1747–1826), and it is today spoken of as Bode's Law. Titius and Bode agreed that there must be, or have been, a body or bodies corresponding to the term $4 + 3 \times 2^3$.

In 1781 came a wholly unexpected development: William Herschel (1738–1822), an amateur observer of whom we shall have much to say, was familiarizing himself with the brighter stars when he came across a 'curious' object that turned out to be the planet we now know as Uranus. When mathematicians were able to determine its orbit, they made the remarkable discovery that its distance from the Sun matched the next term in the sequence, $4 + 3 \times 2^6$. This was enough to convince the court astronomer at Gotha, Baron Franz Xaver von Zach (1754–1832), of the validity of the pattern, and he began to search for a planet corresponding to the term $4 + 3 \times 2^3$. Having no success, in 1800 he held a meeting with a group of friends to discuss how best to proceed, and they divided the zodiac – the region where any planet was likely to be – into 24 zones; each zone was to be assigned to a particular observer whose duty would be to police his zone and look out for any 'star' of no fixed abode.

One of the intended patrolmen was Giuseppe Piazzi (1746–1826) of Palermo Observatory in Sicily. Piazzi was currently at work on a star catalogue, and his careful method of working required him to remeasure the position of each star on a subsequent night. On 1 January 1801, before the invitation from von Zach and company reached him, Piazzi measured the position of an eighth-magnitude 'star'; and when he came to remeasure it, he found that it had moved.

Piazzi was able to track the object for only a few weeks before he lost it in the glare of the Sun, and it was thanks to the emerging mathematical talent of Carl Friedrich Gauss (1777–1855) that it was recovered, by von Zach, at the end of the year. Ceres, as Piazzi called it, matched the missing term in Bode's Law, but it was tiny: Herschel (rightly) thought it smaller even than the Moon. Worse still, three more such objects, also tiny, and also matching the missing term, were discovered in the next few years. Herschel proposed calling the members of this new species of heavenly body 'asteroids'. Wilhelm Olbers (1758–1840), a physician and astronomer also involved in the search for the missing planet, thought they might be fragments of what had once been a full-sized planet.

The search continued for a number of years, but it proved fruitless, and was eventually abandoned. Not until 1845 was another asteroid discovered, by K. L. Hencke, a German ex-postman, and his second success two years later revived interest. By 1891 more than 300 asteroids had been found, and photography was now simplifying the search. Max Wolf (1863–1932) at Heidelberg would photograph a large star field over several hours with a telescope that tracked the rotation of the sky; stars would appear as points of light, but an asteroid would leave a trail as it moved relative to the stars.

Had Olbers been correct, the orbit of each asteroid would – initially at least – have passed through the place where the planet disintegrated, and also through the matching place on the opposite side of the Sun. This proved not to be the case, and astronomers now think that asteroids, whose combined mass is only a fraction of that of the Moon, are objects that failed to coalesce into a planet because of the attractive pull of Jupiter.

The discovery of Uranus had extended the sequence of Bode's Law, but the planet's movements quickly proved puzzling. Early determinations of its orbit were greatly simplified by the discovery that it had been observed (and listed as a star) as long ago as 1690; yet the planet soon began to deviate from its predicted path. Various explanations were proposed, and these were eventually narrowed down to two – either the formulation of the inverse-square law required amendment at such distances, or Uranus was being pulled by a planet as yet undiscovered – and then to one: the undiscovered planet. By the 1840s two talented mathematicians were at work at their desks, putting pen to paper and hoping, by mere calculation, to tell astronomers where to look for this unknown (and previously unsuspected) satellite of the Sun.

The younger of the two was John Couch Adams (1819–92), a graduate of Cambridge, where James Challis (1803–82) was professor. In the autumn of 1845, at Challis's suggestion, Adams visited Greenwich to explain his calculations to the Astronomer Royal, George Biddell Airy (1801–92). By an unlucky chance he failed to see Airy in person, but he left a summary of his results. Next summer Airy was astonished to receive from Paris a copy of a paper by Urbain Jean Joseph Le Verrier (1811–77) predicting the presence of a planet in almost the same position. Airy was of the view that research was not the purpose of the national observatory that he directed, but he asked Challis to institute a search in Cambridge.

The only procedure open to Challis was to plot the star-like objects in a region of sky with care, and then to return to the same region at a later date to see if one of them had moved. This was inevitably a tedious and time-consuming process, and Challis saw no urgency. Unfortunately for the future of Anglo-French relations, Le Verrier had asked astronomers at the Berlin Observatory to make a search, and they – unlike Challis – had a copy of the relevant sheet of the Berlin Academy's new star atlas. They were therefore in a position to compare the stars in the sky with those in the atlas, and within minutes of starting their search on 23 September 1846 they found a star-like object not on the sheet. It was the missing planet.

Challis, it later transpired, had noted the same 'star', but he had not yet returned to the same region of sky to remeasure its position. To the English, Adams's moral claim to the discovery of the planet Neptune was equal to that of Le Verrier, but that was not how the French saw it. But whatever the correct view of this priority dispute, the triumph of Newtonian mechanics was now complete.

This pleasing state of affairs was not to last. Like Uranus, Mercury had an unexplained feature to its orbit: its point of nearest approach to the Sun was advancing in longitude more rapidly than expected, by about one degree per century – a tiny amount, but one that nevertheless called for explanation. Le Verrier, not surprisingly, suspected yet another unseen planet; and in September 1859, he announced that a planet the same size as Mercury but at half the distance from the Sun (and therefore difficult to observe) would be one possible explanation of the phenomenon. It happened that an unknown French physician named Lescarbault had earlier that year seen an object crossing the Sun (or so he thought), and when he read of Le Verrier's prediction, he wrote to him. Le Verrier satisfied himself as to Lescarbault's reliability, and named the planet that the physician had supposedly seen Vulcan. Various alleged sightings of Vulcan followed, but few were convincing; and by the end of the century Vulcan had been rejected as spurious. In 1915 Albert Einstein was to show that the anomalous behaviour of Mercury was implied by his General Theory of Relativity: there was more to the universe than was dreamt of in Newtonian philosophy.

Chapter 6
Exploring the universe of stars

Until 1572, astronomers viewed the 'fixed' stars – fixed and unchanging, that is, in position but also in brightness – as little more than a backdrop to the motions of the planets. In fact, of course, stars do have individual, or 'proper', motions across the sky, but the scale of interstellar distances is so immense that light from even the nearest stars takes years to reach us. As a result, proper motions are almost imperceptible, except over very long time-scales; and so the Renaissance observer saw the stars in positions seemingly no different (except for the overall effect of precession) from those assigned them by Ptolemy.

That no changes in brightness had been noticed is perhaps more surprising. Although most stars, like the Sun, are almost invariable, a minority appear brighter at some times than at others. Some diminish in brightness as they are eclipsed by a companion star, while others are subject to major physical changes, whether regular or irregular. But it so happens that none of these variable stars was bright enough for the variations to force themselves on the attention of an Aristotelian observer of the Middle Ages, convinced as he was that the celestial regions were immutable. Why look for change when you already know that change is impossible?

The result was that it took a wake-up call from Nature – the appearance of the new star, or nova, of 1572 (as seen by Tycho

Brahe, see Chapter 4) – to draw attention to the stars as objects that change and are therefore of interest. Another such nova shone forth in 1604, and this one generated alarm and despondency across Europe. For the first time in eight centuries, the slow-moving planets Jupiter and Saturn were in conjunction in the fateful 'fiery trigon' of the zodiac; and no sooner had they been joined there by Mars, than this new star blazed forth in their midst – the most ominous astrological event imaginable.

No one could now doubt that changes occurred in the heavens. Indeed, there was talk of another nova that had appeared in the constellation of the Whale, but this was fainter, and had been seen by only a single observer before it faded and vanished. In 1638 the Whale was host to a second nova (or so it seemed); like its predecessor, it faded and vanished – but before its discoverer could publish his account of it, it astonished him by reappearing. It continued to vanish and reappear at intervals, and in 1667 Ismael Boulliau (1605–94) announced that this 'wonderful star' was reaching maximum brightness every 11 months: its behaviour was to some extent predictable, and therefore lawlike.

Boulliau went on to offer a physical explanation of variable stars that was ingenious, indeed too much so. He pointed out that variations in sunspots show that the Sun itself – by now recognized as simply the star nearest to us – is, strictly speaking, variable. Furthermore, the rotation of sunspots had demonstrated that the Sun as a whole rotates; and no doubt other stars did likewise. Imagine, then, a rotating star with extensive dark patches instead of mere spots; whenever a dark patch is facing towards us, we shall see the star diminished in brightness, and this will happen regularly, with every rotation of the star. But if the patches themselves vary irregularly, as do sunspots, then this will result in irregular changes in brightness. In this way Boulliau was able to explain both regular and irregular variations. Indeed, so successful was he that the physical explanation of variable stars ceased to be a problem, and astronomers contented themselves with announcing their discovery of variations in particular stars. But these claims could not easily be verified or falsified; the number of allegedly variable stars grew by leaps and bounds, and the whole subject fell into something of disrepute.

Towards the close of the 18th century, the task of verifying such claims was simplified when William Herschel published a series of 'Catalogues of the Comparative Brightness of the Stars'. In these lists Herschel carefully compared stars with neighbouring ones of similar brightness, so that a variation in one of the stars would reveal itself by disturbing the published comparisons. Herschel was generalizing a method involving sequences of stars arranged in order of brightness, which had been developed in the early 1780s by two amateurs who were neighbours in the city of York in the north of England. Edward Pigott (1753–1825) was the son of an accomplished observer; his youthful friend, John Goodricke (1764–86), was a deaf-mute who eagerly accepted the invitation to collaborate in the study of variables.

One of the stars they scrutinized was Algol, which a century earlier had twice been reported as fourth magnitude instead of the usual second. On 7 November 1782 it was second magnitude as usual, but five nights later it was down to fourth; the next night it was back to second. Changes at this rate were unprecedented, so both men now kept watch on the star. On 28 December their efforts were rewarded when they saw Algol start the evening at third or fourth magnitude, and brighten to second before their very eyes. Pigott instantly suspected that Algol was being eclipsed by a satellite, and next day sent Goodricke a note in which he calculated the future orbit of the hypothetical satellite, on the assumption that in the 46 days between 12 November and 28 December, the satellite had completed either one or two orbits. In fact, their observations during the coming months showed that the satellite – if indeed this was the explanation – orbited Algol in less than three days, a phenomenon hitherto unknown to astronomy.

Pigott generously left it to his handicapped friend to make the formal announcement to the Royal Society, but Goodricke, who was still in his teens, merely mentioned the eclipse theory as one possible cause, alongside the traditional dark patches. Pigott had in fact been right – Algol is indeed eclipsed by a companion – but the two friends eventually reverted to the dark patches explanation, perhaps because they wrongly thought they had evidence of irregularities in the light curve of Algol, or because the three other short-period variables that they discovered could not be explained by the eclipse theory. Two, in fact, were cepheids, pulsating stars that climb rapidly to a maximum brightness and then slowly decline, and which would one day be employed as distance indicators by Edwin Hubble and his contemporaries.

The outcome was that 18th-century astronomy yielded a new class of variable stars, those with periods of only a few days, but little progress in understanding the underlying physical causes of these phenomena.

Pigott and Goodricke had seen Algol vary in brightness in a matter of hours. By contrast, changes in position are detectable only in the very long term. Relatively few stars have a proper motion of as much as one second of arc per annum, and the largest known is only just over ten seconds of arc. Such motions can be detected only by comparing the modern position of the star with its position as recorded in a catalogue from an earlier epoch; and, other things being equal, the longer the interval of time since the earlier catalogue, the more accurate will be the resulting value for the proper motion. Unfortunately, however, things are not equal: standards of accuracy decline as we travel backwards in time, and any uncertainty in the position of the star as stated in the earlier catalogue will affect the precision with which its proper motion can be determined.

The only star catalogue from antiquity is that in Ptolemy's *Almagest*; and it was while using this catalogue in 1718 to determine the rate of change of the obliquity of the ecliptic – the angle at which the ecliptic is inclined to the celestial equator – that Edmond Halley realized that three stars must have moved independently of the rest.

As it happened, it would not have been easy for Halley to pursue the question further, for the only other past catalogue of value was Tycho's. This was much more accurate than Ptolemy's; but it was little more than a century old, and its author had dealt crudely with refraction, the bending of starlight as it enters the Earth's atmosphere (which affects the observed position of the star in the sky). Future generations, however, would be able to take the 'British Catalogue', compiled with the greatest care by John Flamsteed at Greenwich, as their starting point in time from which to measure proper motions.

Or so it seemed. But then, in 1728, James Bradley (1693–1762) announced a wholly unexpected complication: the 'aberration of light'. The speed of light is very great, but it is nevertheless finite, as had been shown late the previous century by observation of eclipses of the satellites of Jupiter. These seemed to occur ahead of schedule when the planet was near Earth and the light bearing news of the eclipse had less far to travel, and behind schedule when Jupiter was away on the far side of the Sun.

By comparison, the speed of the Earth in its orbit around the Sun is small, but it is nevertheless big enough to affect the observed positions of the stars. A star appears to an observer to lie in the direction from which its starlight arrives; and this direction alters (slightly) with changes in the direction in which the Earth is moving – just as rain that is in fact falling vertically seems to strike our faces from the direction towards which we are currently moving.

How Bradley came to discover aberration we shall see later in this chapter. The implication of his discovery was that even the British

Catalogue would be seriously defective as the starting point in time for the measurement of proper motions. A further defect became apparent in 1748, when Bradley announced that the Earth's axis 'nutated', or nodded. This was a result of the varying gravitational pulls of the Sun and Moon on an Earth that is not a perfect sphere, and it led to a movement of the very coordinate system used in the measurement of stellar positions.

Bradley himself became Astronomer Royal in 1742, and from 1750, until his health began to fail, he carried out a programme of observations in which he meticulously recorded all the circumstances that might influence the observed position of a star. But he himself did not live to 'reduce' his observations – to make the calculations necessary to derive the true positions of his stars. The reductions had to wait until 1818, when the great German mathematician Friedrich Wilhelm Bessel (1784–1846) published the aptly named *Fundamenta astronomiae*, with over 3,000 stellar positions for 1755, a convenient date midway through Bradley's observing campaign. From then onwards, 19th-century astronomers were able to compare the current position of a star with its 1755 position as given in the *Fundamenta* and so determine how far across the sky the star had moved each year in the interval.

Bradley himself pointed out in 1748 that all proper motions are relative: we do not observe how a star is moving in absolute space, but how it is moving relative to us. Twelve years later, Tobias Mayer discussed the implications of this. If every star except the Sun were at rest, then the motion of the solar system through space would reveal itself to us as a pattern of (apparent) movements among the stars. Therefore, any pattern among the known proper motions is likely to reflect a motion of the solar system; the residual motions will be those of the individual stars themselves.

A modern analogy illustrates the form such a pattern might take. If we drive a car in a city at night, a cluster of distant traffic lights appear bunched together, but they seem to move apart as we approach. At the same time, the street lights on our left appear to move anti-clockwise, while those on our right seem to move clockwise.

Mayer himself could find no such pattern in the (unreliable) proper motions known to him, but in 1783 William Herschel – for once working entirely at his desk – believed he had found a pattern that implied that the solar system is moving towards the constellation Hercules. Today there is no doubt that his conclusion was correct, but his argument does not withstand close scrutiny. A generation later, Bessel had unique access to reliable proper motions in the months during which his *Fundamenta* was in press, and he possessed all the mathematical talent necessary to unravel any pattern; yet he drew a blank.

It was only in 1837 that astronomers became convinced that a solution was in sight. In that year, F. W. A. Argelander (1799–1875), professor of astronomy at Bonn, published an analysis of no fewer than 390 proper motions which he divided by size into three groups, and each group independently yielded a direction not far from that proposed by Herschel.

His conclusions were quickly confirmed by analyses by other astronomers; yet these all depended on the same basic data – Bradley's observations of the stars visible from England. However, Nicholas-Louis de Lacaille (1713–62) had visited the Cape of Good Hope in 1751–53 and determined the positions of nearly 10,000 stars, and 19th-century positions for some of these southern stars were now becoming available. In 1847, 81 resulting proper motions – data that owed nothing whatever to Bradley – were analysed by an actuary, Thomas Galloway (1796–1851), and he derived a direction similar to that based on data for northern stars. Since then there has been no doubt that the solar system is moving in the direction of Hercules, and subsequent analyses of the known proper motions – a list that has increased greatly in number and accuracy with the passage of time – have served only to refine this conclusion.

How far away are the stars? For Ptolemy in antiquity, and for Tycho Brahe in the late 16th century, the fixed stars were just beyond the outermost planet. But if Copernicus was right, then every six months we observe the stars from opposite ends of a huge baseline whose length is twice the radius of the Earth's orbit around the Sun (two 'astronomical units'). As we have seen, even Tycho with his precision instruments was unable to detect the apparent movement among the stars that would result ('annual parallax'), and he very reasonably saw this as refutation of the heliocentric hypothesis.

Part of the problem lay in the nature of the observations: as the months pass, seasonal changes in temperature and humidity will cause an instrument to warp, refraction will vary in response to changes in air pressure, and so on. Galileo, ingenious as ever, saw a way to overcome these difficulties. Suppose two stars lie in almost the same direction from Earth, and suppose one is much further than the other.

The further will have a much smaller parallax than the nearer; this means that we shall not go far wrong if we ignore the parallax of the further altogether, and take it as a quasi-fixed point in the sky from which to measure the parallax of the nearer star. The advantages of this will be great, for the two stars will be equally affected by any warping of the instrument, changes in refraction, and so forth, and the effects of such complications will be eliminated from consideration.

Galileo, indolent as ever, made no attempt to follow up his own suggestion, and it would be many years before it bore fruit. Meanwhile, René Descartes convinced the learned world that stars are suns and the Sun merely our local star, and this suggested a different approach to the problem of stellar distances.

If space is perfectly transparent, light falls off with the square of the distance. Therefore, if the Sun is removed to a thousand times

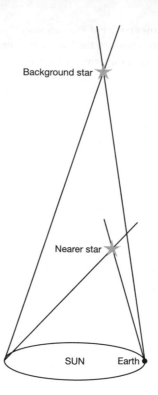

Background star

Nearer star

SUN   Earth

15. Galileo's method of detecting annual parallax, by measuring the apparent annual motion of a nearby star relative to a background star

further than it is at present, it will appear to shine with one-millionth of its present brightness. Suppose now that the stars are not merely of similar nature to the Sun but physically identical to it, so that Sirius (for example) is the physical twin of the Sun. Then if Sirius is found to be one-millionth the brightness of the Sun, and provided space is transparent, we shall know that Sirius is one thousand times further away than the Sun.

But how is one to make the comparison between the glare of the bright Sun and the faint light of a star? The Dutch physicist

Christiaan Huygens (1629–95) put a screen between himself and the Sun, and made a tiny hole in it. His intention was to vary the size of the hole until the portion of the Sun visible through it was equal in brightness to Sirius, and then to calculate what fraction of the Sun was thus visible. It was a crude method, but his result – that Sirius is 27,664 astronomical units from us – was the only estimate available in print for over a quarter of a century after its publication in 1698, and so was widely quoted. Evidently the stars were a long way away.

Meanwhile, unknown to all but a tiny circle of intimates, Isaac Newton had made much better progress, by using an ingenious proposal of the Scottish mathematician James Gregory (1638–75). In a little-noticed book published in 1668, Gregory proposed simplifying the photometric comparison by using a planet as a substitute for Sirius. One was to wait until the planet was equal in brightness to Sirius, and then use one's knowledge of dimensions within the solar system to compare the Sun's light that comes to us directly with the Sun's light that comes to us via the planet. Working on these lines, Newton put Sirius at one million astronomical units. As it happens, Sirius is rather more than half that distance, and so among Newton's intimates, the enormity of the distances that separate the Sun from the nearer stars was now fully appreciated.

But such estimates, based on the provisional assumption that every star is a twin of the Sun, were no substitutes for the actual measurement of the annual parallax (and hence distance) of specific stars. It occurred to Robert Hooke that since the star Gamma Draconis passed directly overhead his lodgings in London, its light would then be unaffected by atmospheric refraction. He tried to avoid the danger of seasonal warping of the observing instrument by incorporating the components of a telescope into the actual fabric of his house. Although telescopic astronomy was in its infancy, Hooke designed and built a telescope to observe one single star, at just one moment during its passage, and for one purpose only.

Hooke's ingenuity was not matched by his perseverance: he made a mere four observations, in 1669, before illness and an accident to the telescope lens brought his efforts to an end. But his method had much to commend it, and in the mid-1720s a prosperous English amateur, Samuel Molyneux (1689–1728), decided to make another attempt to measure the annual parallax of Gamma Draconis. He invited James Bradley to join him, and he commissioned a special 'zenith sector' from a leading maker, George Graham. The sector with its vertical telescope was mounted on a chimney stack in Molyneux's house, and when a star passed overhead the tube of the telescope was tilted slightly to bring the star into the middle of the field of view; the angle of tilt was measured against the scale of the sector to give the angular distance of the star from the vertical.

A simple calculation showed that Gamma Draconis should reach an extreme southerly position a week before Christmas, and so it was a surprised Bradley who on 21 December saw it pass overhead markedly further south than it had a week earlier. By March, when it should have been moving north, it was some 20 seconds of arc south of its December position. The star then stopped and back-tracked, passing through its December position in June, and reaching an extreme northerly position in September.

The two friends debated various explanations – was there a movement in the axis of the Earth and therefore in the coordinate system by which the star's position was being measured, or was the atmosphere of the Earth distorted by the planet's passage through space so that atmospheric refraction was unexpectedly affecting the measures? – but without success. Bradley decided to commission from Graham another zenith sector, this time with a wider field of view that would bring more stars under scrutiny, and with it he established the patterns of the stellar movements; but their explanation eluded him. Then, one day, while on a boat on the Thames, he noticed that the weather vane changed direction as the boat put about – not of course because the wind had veered, but because the boat had altered course. The starlight, he now realized, was likewise reaching the observer from changing directions, because the observer was altering course as the Earth orbited the Sun.

This discovery of aberration, announced to the Royal Society in 1729, was momentous, for several reasons. It was the first direct proof of the Earth's motion around the Sun. Because all stars were similarly affected, it showed that the velocity of light was a constant of nature. It revealed (as we saw earlier) a wholly unexpected error in past measurements of stellar positions, Flamsteed's among them. And because even Bradley's zenith sector, for all its accuracy, was unable to detect annual parallax, the stars must lie at distances of at least 400,000 astronomical units.

Only the previous year, the posthumous publication of Newton's *The System of the World* had made public his estimate – based, it is true, on the working hypothesis of physical uniformity among the stars – that Sirius lay at one million astronomical units. These two results – the one giving an actual distance but on the basis of a questionable hypothesis, the other a minimum distance based on direct measures – combined to convince astronomers that the scale of stellar distances was at last understood.

The unwelcome implication was that annual parallax must be at most a second or two of arc, an angle so tiny as to represent the width of a coin at a distance of kilometres. Such a minute movement, taking place over a period of months, would be well-nigh impossible to detect, and the next generation of astronomers displayed little enthusiasm for such a hopeless task. William Herschel in the 1770s and 1780s collected a great number of double stars, ostensibly for use in Galileo's method of measuring parallax; but he was playing the natural historian, collecting specimens that others might one day put to use. Most astronomers preferred to spend their time on more promising lines of enquiry.

In any case, John Michell (c.1724–93) had shown in 1767 – unknown to Herschel – that the number of double stars was so large that most must be true companions in space (binary stars), lying at the same distance from the observer, and therefore useless for Galileo's method of measuring parallax. Herschel himself was to confirm Michell's claim when he re-examined some of his doubles around the turn of the century, and found instances where the two stars had orbited relative to each other. A generation later William's son John was to be one of those who confirmed that their orbits were indeed ellipses, and that the force binding the companion stars together was therefore Newtonian attraction. Newton had claimed that attraction was a universal law, but this was the first evidence that the law applied outside the solar system.

Meanwhile, astronomers found themselves in a situation where, as telescopes improved, the two coordinates of a star's position on the heavenly sphere were being measured with ever increasing accuracy, whereas little was known of the star's third coordinate, distance, except that its scale was enormous. Even the assumption that the nearest stars were the brightest was being called into question, as the number of known proper motions increased and it emerged that not all the fastest-moving stars were bright.

An extreme example of this was found early in the 19th century when first Piazzi and then Bessel found that the relatively faint star 61 Cygni was travelling across the sky with the exceptional speed of over 5 seconds of arc per annum. Surely this showed that the star must be near, despite its modest brightness?

Annual parallax is of course inversely proportional to distance, and it was crucial that observers attempting to measure this parallax concentrate their efforts on the stars closest to Earth. In 1837, after several claims to success in measurement had proved ill-founded, the German-born Wilhelm Struve (1793–1864) proposed three criteria of nearness: was the star bright, was its proper motion sizeable, and – if it happened to be a binary star – did the two components appear widely separated in relation to the time they took to orbit each other?

At his observatory at Dorpat (now Tartu in Estonia) Struve was privileged to possess a magnificent refracting telescope by Joseph Fraunhofer (1787–1826). Its object glass was no less than 24 cm in diameter and of exceptional quality, and its mounting was 'equatorial', its axis pointing to the north celestial pole so that the observer needed to rotate only this one axis to keep the telescope aligned on a star. In 1835 Struve had selected for his parallax measurements the star Vega, which is very bright and has a large proper motion, and in 1837 he announced the results of 17 observations, from which he inferred a parallax of one-eighth of a second of arc. Three years later he reported on 100 observations, and this time inferred a parallax of one-quarter of a second. But given the long history of spurious claims, going back to Hooke, astronomers had yet to be convinced.

Meanwhile, Bessel, at Königsberg, was equally fortunate in his instrumentation. His Fraunhofer refractor was not as large, the objective having a diameter of just 16 cm. But its maker, not content with achieving a lens of high quality, had then taken his courage in both hands and cut it into two semicircular pieces of glass that could move against each other along their common diameter. Each half showed a complete image, but of only half the brightness. If the telescope was turned towards a double star, the pair of stars would appear in both halves, and the observer could then slide the halves relative to each other until one star in one image coincided with the other star in the other image. The amount of displacement necessary was then a very accurate indication of the angle separating the two stars. Because such instruments were often used to monitor changes in the apparent diameter of the Sun, they were known as heliometers.

Bessel selected for scrutiny 61 Cygni, known as the 'flying star' because of its large proper motion. In 1837 he subjected the star to an unprecedented examination, observing it for over a year as many as 16 times in a single night – even more often if the 'seeing' was particularly good. The following year he was able to announce that it had a parallax of about one-third of a second of arc. What carried conviction was the way in which the plotted graph of his observations matched the expected theoretical curve. It was, John Herschel told the Royal Astronomical Society, 'the greatest and most glorious triumph which practical astronomy has ever witnessed'. The stellar universe now had a third dimension; and the number of successful measurements of annual parallax would multiply in the decades to come.

But what was the large-scale structure of this universe? Newton's *Principia* had almost nothing to say about the stars, and it seems that he had given questions of cosmology little thought prior to 1692, when he received a letter from a young theologian, Richard Bentley (1662–1742). Bentley had given a series of lecture/sermons on science and religion, and before putting these into print he wanted to know the views of the author of that densely mathematical book that everyone respected but no one understood. Bentley had no time for the Cartesian position whereby God had created the universe and left it to run itself; but he wanted to know what could be said in support of this, and so he asked Newton what would happen in a universe in which the matter was initially distributed with perfect symmetry. Newton, not realizing that Bentley intended the symmetry to be literally perfect, replied that wherever the matter was more concentrated than usual, its gravitational pull would attract the surrounding matter and so lead to still greater concentration. Corrected, and irritated, he conceded that in a *perfectly* symmetric universe the matter would have no reason to move one way rather than another; but he commented that a perfect symmetry was as plausible as having infinitely many needles all standing on their points on an infinite mirror. 'Is it not as hard', retorted Bentley, 'that infinite such Masses in an infinite space should maintain an equilibrium?' In other words, how was it that the stars were all 'fixed' and motionless, when each was supposedly pulled by the gravitational attraction of all the others?

Newton was now squarely confronted with the paradox underlying the claim in the *Principia* that attraction is a universal law of nature; for it seemed that, even after observations of centuries, the stars were as fixed as ever. Curiously, Newton was (as we have seen) the only person to have a correct appreciation of the scale of interstellar distances; yet it did not occur to him that, because the stars were so enormously far away, any motions they had would be well-nigh imperceptible. Instead, he continued to believe that the stars – the *fixae* – were motionless, and his problem was to explain how this could be so.

His solution to the paradox is to be found in drafts for an intended second edition of the *Principia* that he abandoned when he left Cambridge for a post in London. We remember that he saw the finite system of planets orbiting the Sun as having been planned by Providence to provide a stable environment for mankind, though the stability was not perfect and therefore Providence would eventually have to intervene, to prevent gravity from undermining the system. The system of the stars was likewise stable; but this, he argued, was because the stars were infinite in number and their distribution (almost) symmetric: each star was initially at rest, and it remained so because it was pulled equally in every direction by the other stars.

Yet a glance at the night sky would show that the symmetry was not in fact perfect; and indeed Newton needed ingenuity to provide evidence for at least the semblance of symmetry among the nearer stars. But he did not see imperfect symmetry as a problem: instead, it was another area for regular interventions by Providence, as a result of which the stars would be restored to their earlier order.

Newton had focused on the dynamics of the stellar universe, but what of the light sent to us by these same stars? This question was put to Newton around 1720 by a young physician of his acquaintance, William Stukeley (1687–1765). Galileo's telescope, a century before, had confirmed that the Milky Way resulted from the combined light of innumerable tiny stars; but – curiously – there had been little subsequent interest in the three-dimensional distribution of stars that gives rise to this phenomenon, and it did not occur to Newton that the Milky Way was disproof of his claim that the universe of stars is symmetric.

Stukeley, however, speculated that the stars bright enough to be individually visible might together form a spherical assemblage, while the stars of the Milky Way formed a flattened ring surrounding this sphere – in effect, a stellar analogue to Saturn and its ring. In response Newton hinted that an infinite symmetric universe of stars was to be preferred; to which Stukeley – not knowing that Newton was secretly committed to this very concept – retorted that in such a universe 'the whole hemisphere [of the sky] would have had the appearance of that luminous gloom of the milky way'.

Early in 1721, Stukeley and Halley took breakfast with Newton, and they discussed questions of astronomy. These must surely have included the possibility of an infinite universe of stars, for a few days later Halley read to the Royal Society the first of two papers on the subject. When his papers were published in *Philosophical Transactions*, Newton's model of the universe came at last – if anonymously – into the public domain.

In one of the papers Halley remarked carefully: 'Another Argument I have heard urged, that if the number of Fixt Stars were more than finite, the whole superficies of their apparent Sphere would be luminous.' He had his own solution to Stukeley's misgivings, but it was flawed. It was not until 1744 that a correct analysis of light in an infinite and nearly symmetric universe was published. The Swiss astronomer J.-P. L. de Chéseaux (1718–51), pointed out that at the distance of the nearest stars there was (so to speak) room for a given number of stars such that no two were unduly close together; and that in aggregate these stars filled a certain (tiny) area of celestial sphere. At twice the distance there was room for four times as many stars, but each would be one-quarter the brightness and one-quarter the apparent size. In aggregate, therefore, they would fill the same area of the sky as before, and to the same level of brightness. At three times the distance, the stars would fill a further area of sky with light; and similarly for each succeeding step, until eventually the entire sky was ablaze with light.

Or so one might think (and indeed modern astronomers see the darkness of the night sky as posing 'Olbers's Paradox'). But Chéseaux pointed out – as did Olbers in 1823 – that this reasoning assumes that all the light that sets out from a given star reaches its destination; whereas even a minute loss of light, repeated at each stage of the journey, would effectively reduce the very distant stars to invisibility. For neither Chéseaux nor Olbers was there any paradox.

Nor was there for later 19th-century astronomers, even though by now it was realized that an interstellar medium that intercepted light would itself heat up and begin to radiate. There were plenty of other ways out of the difficulty, such as the existence of etherless vacua across which no light could pass. Only in our own time has the darkness of the night sky been elevated to the status of a paradox, and those who christened it were unaware that the question goes back beyond Olbers, to Chéseaux, Halley, and ultimately to the physician Stukeley.

Meanwhile, amateur speculators began to puzzle over the Milky Way. In 1734 Thomas Wright of Durham (1711–86) gave a public lecture/sermon in which he presented his highly personal cosmology. The Sun and the other stars, he told his audience, orbited around the Divine Centre of the universe, occupying as they did so a spherical shell of space outside of which was the Outer Darkness; and he concentrated the minds of his audience by

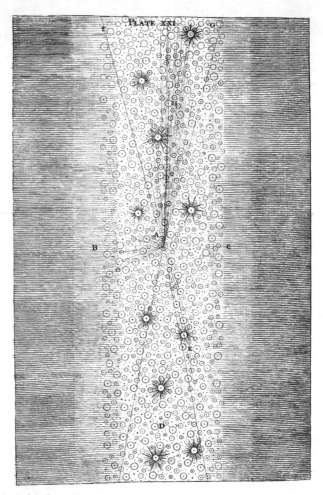

16. A sketch used by Wright to help readers understand his preferred model of our system of stars. In this imaginary universe there is a layer of stars bounded by two parallel planes. An observer at *A* would see only a handful of near (and therefore bright) stars when looking outwards from the layer, in the direction of *B* or *C*; but when looking along the layer, in the directions of *D, E*, etc., the observer would see innumerable stars, near and far, whose light would merge to give a milky effect. From Thomas Wright, *An Original Theory of the Universe* (1750).

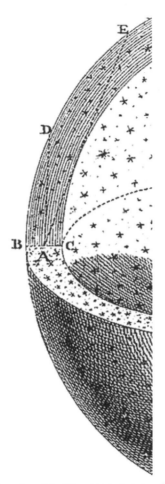

17. Wright's preferred model of the star system to which the solar system belongs. The space occupied by the stars has the form of a spherical shell, whose radius is so vast that its curvature is imperceptible to the human observer located at *A*. For the observer, therefore, the inner and outer surfaces of the layer of visible stars approximate to parallel planes. As before, the observer sees only a handful of near and therefore bright stars when looking in the direction of *B* or *C*, but sees innumerable stars whose light merges to give a milky effect when looking along the layer, in directions such as *D* and *E*.

pointing out that each of them was destined, after death, to pass either inwards or outwards. To force this home, he prepared a visual aid displaying a cross-section of the universe, and in it, by artistic licence, he portrayed the solar system and the visible stars as they actually appear from Earth; the light of the more distant of these stars, he said, merged to form 'a faint circle of light', the Milky Way.

It was only later than he realized his mistake: such a Milky Way would lie in every single cross-section of the universe that passed through both the Divine Centre and the solar system, whereas the real Milky Way is unique. In his beautifully illustrated *An Original Theory or New Hypothesis of the Universe*, published in 1750, he met this difficulty by greatly reducing the thickness of the spherical shell of space occupied by the Sun and the other stars of our system (for now he envisaged many such systems, each with its own Divine Centre). As a result, when looking inwards or outwards we see only a few, near (and therefore bright) stars, before our gaze extends into empty space. But when we look tangentially to the shell, whose radius is vast and which therefore curves imperceptibly, we see great numbers of stars whose light merges to create a milky effect: that is, the plane of the Milky Way is tangent to the shell at our particular location.

A summary of Wright's book, but without the illustrations necessary to comprehend his bizarre conceptions, appeared in a Hamburg periodical the following year and came to the attention of the German philosopher Immanuel Kant (1724–1804). Kant assumed, not unreasonably, that there must be only one Divine Centre, away in some remote part of the universe, and that the region of our star system was therefore entirely in the natural order. He knew of milky patches (nebulae) that had been observed in the sky and which he believed to be other star systems; but these were elliptical, whereas a spherical system seen from without will always appear circular. Kant therefore opted for an alternative model offered by Wright, in which the stars surrounding our Divine Centre formed a flattened ring. As far as Kant was concerned, there seemed no reason why this ring of stars (being wholly in the natural order) should not extend without a break from one side to the other, thus forming a complete disc of stars; and a disc seen edge-on will appear elliptical, just like the nebulae that had been observed. Kant therefore credited Wright – mistakenly – with the conception of the Milky Way as a disc-shaped aggregate of stars, which indeed it is.

These and similar speculations, dreamt up by amateurs in the warmth of the study, were hardly likely to impinge on professional astronomers. On the other hand they could scarcely ignore the discovery of the planet Uranus in 1781 by another amateur, the familiar William Herschel, a musician who had come to England from Hanover as a refugee from the Seven Years' War. But the amateurishness of his report, and his casual claim to have made the discovery with eyepieces whose alleged magnifications were beyond the powers even of professional opticians, made him a figure of controversy.

In 1772 Herschel had rescued his sister Caroline from family servitude in Hanover, and she was to be his devoted assistant in everything he did. His enthusiasm for astronomy soon began to take over their lives, Herschel's ambition being nothing less than to understand 'the construction of the heavens'. Herschel realized that to see objects that were distant and therefore faint, he must equip himself with reflecting telescopes able to collect as much light as possible – in other words, with the largest possible mirrors. He learned to grind and polish discs purchased from local foundries, but his ambitions soon outreached their capacities: a 3-foot disc he would have to cast himself. Undaunted, in 1781 he turned the basement of his own home into a foundry, but his two attempts resulted in failure and near-tragedy.

The discovery of Uranus gave Herschel's admirers the opportunity to lobby the King on his behalf, and in 1782 Herschel was awarded a royal pension that enabled him to devote himself to astronomy. He moved to the vicinity of Windsor Castle, where he had no duties

18. William Herschel's 'large' 20-foot reflector, commissioned in 1783, from an engraving published in 1794. His 'sweeps' with this instrument resulted in the discovery of 2,500 nebulae and clusters of stars. By 1820 the woodwork was in an advanced state of decay, and William's son John was forced to build a replacement, which he took to the Cape of Good Hope to extend his father's work to the southern skies.

except to show the heavens to the royal family and their guests when asked. He soon constructed one of the great telescopes of history, a reflector of 20-foot focal length and mirrors 18 inches in diameter and, equally important, a stable platform.

With Caroline seated at a desk within earshot, ready to act as amanuensis, Herschel devoted much of his observing time over the next two decades to 'sweeping' the night sky for nebulae and clusters of stars. The telescope, facing south, would be set to a given elevation, and alternately raised and lowered a little from this position as the heavens rotated slowly overhead; in this way a strip of sky would be 'swept' for any nebulae it might contain. When they began, only a hundred or so of these mysterious objects were known; when they finished, they had collected and classified 2,500 specimens.

Everyone recognized that a star cluster so distant that the individual stars could not be distinguished would appear nebulous, as indeed did the Milky Way. But were all nebulae distant clusters, or were some formed of nearby clouds of luminous fluid ('true nebulosity', as Herschel termed it)? If a nebula were seen visibly to alter shape, this would prove it was a nearby cloud, for a distant cluster would be too vast to change so rapidly; and in 1774, on the very first page of his first observing book, Herschel had noted that the great nebula in Orion was not as it had been depicted earlier (by Huygens in the 17th century). Occasional observations of the same nebula in subsequent years persuaded him that it was continuing to change, and was therefore formed of true nebulosity. But how to distinguish true nebulosity from a distant star cluster? It seemed to Herschel that he was encountering two kinds of nebulosity, milky and mottled, and he supposed that mottled nebulosity reflected the presence of innumerable stars.

But then, in 1785, he came across a nebula that contained individual stars together with both kinds of nebulosity; and he interpreted this as a star system that extended away from the observer. The nearest stars were individually visible, the more distant appeared as mottled nebulosity, while the most distant appeared as milky nebulosity. Herschel therefore reversed his earlier position, and decided that all nebulae were star clusters.

But clusters implied clustering: an attractive force or forces – presumbly Newtonian gravity – that was operating to pull the member stars together ever more closely. This implied that in the past, the stars of a cluster had been more scattered than they were now; in the future, they would be more tightly clustered.

In this way Herschel was introducing into astronomy concepts from biology: he was acting the natural historian, collecting and

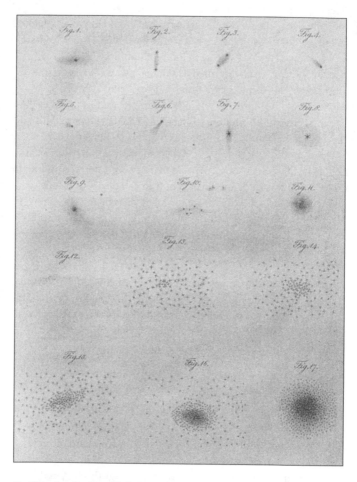

19. Sketches by Herschel showing objects from his catalogues of nebulae and clusters of stars, arranged in order of increasing maturity: gravitational attraction operates to concentrate the clusters more and more as time passes. From *Philosophical Transactions*, vol. 104 (1814).

classifying great numbers of specimens, which he could arrange by age – as young, in middle life, and old. He was changing the very nature of the science.

One evening in 1790, he was sweeping as usual when he came across a star surrounded by a halo of nebulosity. The star, he admitted, must be condensing out of the nebulosity, and so true nebulosity must exist after all. He would have to extend his theory of the development of star systems back in time, to accommodate an earlier phase during which thinly scattered light condensed under gravity into clouds of nebulosity, out of which stars were born. These stars formed clusters, scattered at first but increasingly condensed – until at length the clusters collapsed upon themselves in a great celestial explosion, the light from which began the cycle over again. Herschel's contemporaries, almost none of whom had instruments with which to view the evidence, were at a loss to know what to make of it all.

It was William Herschel's son John (1792–1871) who brought stellar astronomy into the scientific mainstream. When he was a young man his father prevailed upon him to abandon a career at Cambridge and return home, to become in effect his apprentice and astronomical heir, charged with the re-examination and extension of his father's collections of astronomical specimens. William's 20-foot reflector was now decayed with age, but before he died in 1822, he supervised John in the construction of a replacement.

In 1825 John Herschel began the revision of his father's catalogues of nebulae visible from England. This done – and resolutely declining all offers of government help – he set sail for the Cape of Good Hope, where he spent four years extending to the southern skies his father's catalogues of nebulae, double stars, and so forth. He became, and remains, the only observer to have examined the entire celestial sphere with a major telescope.

When John Herschel set sail for home in March 1838, his career as an observer was ended, and so was the Herschelian monopoly on great telescopes. That year, at Birr Castle in central Ireland, William Parsons (1800–67), the future Earl of Rosse, manufactured and assembled the segments to form a composite mirror 3 feet in diameter. The following year he succeeded in casting a single disc of this size, and in 1845 he completed the 'Leviathan of Parsonstown', a monster reflector slung between huge masonry walls, with mirrors no less than 6 feet in diameter and weighing four tons each. Within weeks the reflector had revealed that some nebulae are spiral in structure.

The Leviathan was designed to settle once and for all the question of whether all nebulae were merely star clusters disguised by distance; and all agreed that examination of the great nebula in Orion, which is visible to the naked eye, would be decisive. It so

20. Lord Rosse's reflector with 6-foot mirrors, commissioned in 1845. By April of that year, Rosse had used it to discover the spiral structure of some nebulae.

happened that the reflector was powerful enough to detect stars that are indeed embedded in this (gaseous) nebula, and on seeing these Rosse persuaded himself that he was viewing a cluster that he had triumphantly 'resolved' into its component stars.

Many agreed that this success with the greatest of the nebulae could be generalized, and that belief in 'true nebulosity' had been banished from astronomy for ever. They were soon to be proved wrong, but only after astronomy had surrendered its autonomy and merged with physics and chemistry to embark on the analysis of starlight.

Epilogue

In this book we have followed attempts of observers and theoreticians down the ages to understand the heavenly bodies – what they are, and how they behave. Whether observers consciously thought of it or not, their information came from the light currently arriving at Earth from these bodies: it was light, rather than the bodies themselves, that they observed.

Not all such light was the same. Some stars shone with a brilliant whiteness, for example, while others had a reddish hue. The relationship between colours and the white light that arrives from our nearest star, the Sun, was established in 1666 by Isaac Newton. He made a hole in the shutters of his room in Trinity College, Cambridge, and passed the beam of sunlight through a prism. He saw, as expected, the familiar spectrum with all the colours of the rainbow. The received theory was that white light was simple and basic, and that a colour resulted from some sort of modification of white light: you started with white light, and you did something to it to obtain a colour. By careful experimentation Newton found that, to the contrary, it was the colours that were basic, and that when recombined together they formed white light once more. Sunlight was made up of the colours of the rainbow.

Newton was investigating light itself, and not the Sun as source of the light. William Herschel was the first observer with the curiosity to examine the spectra of light from other stars, and with telescopes of sufficient 'light-gathering power' to make this a practical possibility. As early as 1783 he several times put a prism at the eyepiece of one or other of his 20-foot reflectors when it was directed to a bright star, but it was not until 9 April 1798 that he undertook a brief investigation into the light from six of the brightest stars. 'The light of Sirius', he found, 'consists of red, orange, yellow, green, blue, purple, and violet'. On the other hand, 'Arcturus contains more red and orange and less yellow in proportion than Sirius'. And so on. But what these differences implied he had no idea.

The answer was gradually to emerge from a much more careful analysis of sunlight. In 1802 William Hyde Wollaston (1766–1828) repeated Newton's experiment, but he replaced the crude hole that Newton had made in his shutters with a narrow slit only one-twentieth of an inch wide. He was surprised to find the spectrum of sunlight was crossed by seven dark lines, which he took to be the divisions between the colours. However, when the telescope-maker Joseph Fraunhofer was making tests on glass lenses, he was astonished to discover that there were in fact hundreds of such lines. He also found that spectra that were quite different in form could be generated in the laboratory, consisting of thin bright lines with dark spaces between them (a 'bright-line' spectrum, in contrast to the continuous spectra of the Sun and stars).

Over the next three decades, the situation was gradually clarified, and the far-reaching implications became apparent. Two Germans, the chemist Wilhelm Bunsen (1811–99) and the physicist Gustav Robert Kirchhoff (1824–87), played a central role in this. By 1859 they had established that glowing solids and liquids produced a continuous spectrum, that of sunlight being the familiar example, while glowing gases produced a bright-line spectrum. (As a result, when the English astronomer William Huggins (1824–1910) in 1864 managed to obtain a visible, bright-line spectrum from a nebula in Draco, he ended the centuries-old debate as to whether 'true' (gaseous) nebulae exist.) Each element had its own characteristic line positions. Surprisingly, a continuous spectrum when passed through a gas displayed a 'dark-line' spectrum, with dark lines characteristic of the gas. As a result, once the line positions of an element had been established in the laboratory, the investigator could demonstrate the presence or absence of the element in the star or nebula, or in any gas through which the light from the celestial body had passed.

A Pandora's box was opened. As the great American observer James Keeler remarked, 'The light which reveals to us the existence of the heavenly bodies also bears the secret of their constitution and physical condition'. The limitation on human knowledge famously declared in 1835 by Auguste Comte, that we can never by any means investigate the chemical composition of celestial bodies, was sensationally disproved. So profound was the transformation that astronomy lost its identity, to become a branch of physics (and chemistry). As Huggins put it,

> Then it was that an astronomical observatory began, for the first time, to take on the appearance of a laboratory. Primary batteries, giving forth noxious gases, were arranged outside one of the windows; a large induction coil stod mounted on a stand on wheels so as to follow the positions of the eye-end of the telescope, together with a battery of Leiden jars; shelves with Bunsen burners, vacuum tubes, and bottles of chemicals, especially of specimens of pure metals, lined its walls.

The study of the properties, constitution, and evolution of heavenly bodies became the province of 'astrophysicists', rather than astronomers, and Kepler's title, 'the new astronomy', was invoked once more. Meanwhile, traditional astronomy flourished and developed, in tandem with astrophysics.

The history of any science is never-ending, and the scope of the discipline grows while the number of practitioners escalates. The transformation of astronomy in the mid-19th century marks the end of the story we have set out to tell, and – as always – the beginning of another.

Astronomical research is now the work of teams of scientists and engineers. Radio telescopes intercept radiation – information – at wavelengths imperceptible to the human eye, and combinations of such telescopes may be equivalent to a single 'dish' hundreds of kilometres in diameter. Optical telescopes with ever-larger mirrors are being built on the tops of mountains above most of the Earth's atmosphere, and predominantly in the southern hemisphere where many of the most significant deep-sky objects are to be seen. Computers drive the telescopes, and by 'active optics' continuously compensate for subtle changes in the balance of the mirror and in the atmosphere above the instrument. New technology has enormously increased the amount of incoming information that can be secured from some of the most dramatic of contemporary instrumentation, including the Hubble Space Telescope and the planetary probes. Viewing the images these spacecraft transmit to Earth by radio links makes it easy to understand why there has never been a more exciting time to be an astronomer.

Further reading

The present work is in effect an introduction to two closely related books by the author and colleagues, either of which would serve as a text for further reading. They are: Michael Hoskin (ed.), *The Cambridge Illustrated History of Astronomy* (hereafter *CIHA*; Cambridge, 1997); and Michael Hoskin (ed.), *The Cambridge Concise History of Astronomy* (*CCHA*; Cambridge, 1999). The *Illustrated History* has numerous illustrations in colour, while in the *Concise History* the text (despite the book's title) is amplified with additional technical material. The subjects of our first four chapters are further discussed in articles in Christopher Walker (ed.), *Astronomy Before the Telescope* (*ABT*; London, 1996). The relevant sections of these works are given first in the suggested further reading below. For an alternative overview of the whole history of astronomy, see the paperback by John North, *The Fontana History of Astronomy and Cosmology* (London, 1994).

All the books cited above include bibliographies. Individual astronomers are treated authoritatively in the multi–volume *Dictionary of Scientific Biography*, edited by C. C. Gillispie (New York, 1970–90), available in many reference libraries.

Those wishing to keep abreast of current work in the field may consult the *Journal for the History of Astronomy* (Science History Publications, Cambridge).

Chapter 1

CIHA or *CCHA*, Chapter 1; *ABT*, article by Ruggles.

The customs of orienting buildings on heavenly bodies in prehistoric Europe and the Mediterranean area are discussed in Michael Hoskin, *Tombs, Temples and Their Orientations: A New Perspective on Mediterranean Prehistory* (Bognor Regis, 2001). For the British Isles, see Clive Ruggles, *Astronomy in Prehistoric Britain and Ireland* (New Haven and London, 1999), a more technical work with discussions of methodology.

Chapter 2

CIHA or *CCHA*, Chapter 2; *ABT*, articles by Wells, Britton and Walker, Toomer, Jones, and Pingree.

A wide-ranging and user-friendly book is James Evans, *The History & Practice of Ancient Astronomy* (New York and Oxford, 1998). Otto Neugebauer, *Exact Sciences in Antiquity*, 2nd edn. (Providence, RI, 1957), is somewhat dated, but the work of a master. Astrology was a powerful motivation for astronomy in antiquity; the best account is Tamsyn Barton, *Ancient Astrology* (London and New York, 1994).

Chapter 3

CIHA or *CCHA*, Chapters 3 and 4; *ABT*, articles by Field, King, and Pedersen.

On astronomy in Christendom, see Stephen C. McCluskey, *Astronomers and Cultures in Early Medieval Europe* (Cambridge, 1998), and Edward Grant, *Planets, Stars, and Orbs: The Medieval Cosmos, 1200–1687* (Cambridge, 1994).

Chapter 4

CIHA or *CCHA*, Chapter 5; *ABT*, articles by Swerdlow and Turner.

Astronomy of the period is treated systematically in *The General History of Astronomy*, Vol. 2: *Planetary Astronomy from the*

Renaissance to the Rise of Astrophysics, edited by R. Taton and C. Wilson, Part A: *Tycho Brahe to Newton* (Cambridge, 1989).

Chapter 5

CIHA or *CCHA*, Chapter 6.

The General History of Astronomy, Vol. 2, Part A, Chap. 13 is the best introduction to Newton's *Principia*, while Part B of the same work, *The Eighteenth and Nineteenth Centuries* (Cambridge, 1995) is excellent on the implementation of the Newtonian programme.

Chapter 6

CIHA or *CCHA*, Chapter 7

Michael Hoskin, *Stellar Astronomy: Historical Studies* (Cambridge, 1982: Science History Publications, 16 Rutherford Road, Cambridge CB2 2HH).

Glossary

aberration of light: the small, constantly varying shift in the observed position of a star caused by the velocity of the Earth-based observer's orbit around the Sun.

annual parallax: the small, constantly varying displacement in the observed position of a star, caused by the displacement of the Earth-based observer from the centre of the solar system.

astronomical unit: the mean distance of the Earth from the Sun (about 93 million miles).

atmospheric refraction: the bending of the path of light from a celestial body by the Earth's atmosphere.

binary star: two stars that are companions in space, bound to each other by their mutual attraction and each in orbit about their common centre of gravity.

bright-line spectrum: see *spectrum* below.

celestial equator: the projection onto the sky of the equator on Earth.

celestial poles: the projection onto the sky of the axis of rotation of the Earth.

centrifugal force: the tendency of a body in orbit to 'fly off at a tangent'.

centripetal force: a force such as gravity that causes a body to move 'inwards' in a curved path, instead of continuing in a straight line.

dark line spectrum: see *spectrum* below.

deferent: in ancient or medieval planetary theory, the circle in the geometric model of a planet's orbit about the Earth that 'carries' either the Sun or the centre of the little circle or 'epicycle' on which the planet is imagined as located.

eccentric circle: in ancient or medieval planetary theory, a circle that is not centred on the Earth.

ellipse: a closed curve formed by the intersection of a cone with a plane; as Kepler showed, the planets orbit the Sun in elliptical paths.

empty focus: as a planet orbits the Sun in its elliptical path, the Sun is located at one of the two foci (a geometrically significant position on the major axis of the ellipse, to one side of the centre); the 'empty' focus is the matching position on the other side of the centre, and is so called because no physical body is located there.

epicycle: in ancient or medieval planetary theory, the small circle that carries the planet in the geometrical model of a planet's orbit about the Earth, and which is itself carried on the deferent.

equant point: in the geometrical model of a planet's orbit in Ptolemy's *Almagest*, the position symmetrically opposite the eccentric Earth; the planet is imagined moving with a (variable) speed such that, viewed from the equant point, its motion across the sky appears uniform.

heliacal rising: the reappearance in the dawn sky of a star or planet after several weeks of invisibility lost in the glare of the Sun.

inverse-square force: a force, such as gravity, that reduces as distance increases, in proportion to the square of the distance.

nebula: a milky patch seen in the sky and so quite different in appearance from a star or planet; physically, nebulae are of several different kinds, some being vast star systems so far away that it is difficult to distinguish the individual stars, while others are wholly or partly gaseous.

nova: *nova stella*, a 'new' star that appears where none was visible before.

Olbers's Paradox: if the stars were scattered at regular intervals throughout infinite space, analysis suggests that the whole of the sky would appear as bright as the Sun; modern cosmologists have erroneously supposed that the fact that this does not happen was seen as a paradox by H. W. M. Olbers (1758–1840).

period (of a planet): the time taken by a planet to complete one orbit of the Sun.

precession of the equinoxes: the slow change of direction of the Earth's axis of rotation (period 25,800 years), as a result of which the celestial equator moves relative to the ecliptic (the path of the Sun), and their points of intersection—the 'equinoxes'—therefore move or 'precess'.

proper motion (of stars): the observed motion of an individual star on the celestial sphere (and so at right angles to the observer's line of sight).

rectilinear inertia: the tendency of a moving body to continue moving in a straight line with uniform speed.

retrograde: the occasional 'backwards' motion of a planet as seen from Earth; for example, the motion of Mars, Jupiter or Saturn appears retrograde for a time when the Earth overtakes it 'on the inside', as the planets together orbit the Sun.

spectrum: the continuous band of 'colours of the rainbow', from violet to red, into which white light is dispersed when passed through a prism. White light is emitted by very hot solids, liquids, or dense gases. A *dark line spectrum* is crossed by transverse dark lines which are formed by cooler gases in the line of sight, e.g. in the cooler atmosphere of the Sun. A *bright line spectrum* has luminous lines on a fainter or dark background, which arise from a hot but highly rarified gas, e.g. in a gaseous nebula.

variable star: a star whose apparent brightness varies, either regularly or irregularly.

Index

A

aberration of light 82
Adams, J. C. 75, 76
Airy, G. B. 75–6
al-Khwarizmi 28
al-Sufi 25
al-Tusi 25, 29
al-Zarqali 32
Alexandria 18
Alfonsine Tables 33, 35, 36, 43
Alfonso X (King) 33
Algol 80–1
Alhazen 29
Almagest 18, 21, 22, 23, 24, 32, 36
Alpetragius 28
annual parallax 38, 42, 49, 55, 64, 85, 89, 92
Apollonius of Perga 15, 18
Aquinas, Thomas 32
Argelander, F. W. A. 84
Aristotle 9, 10–11, 13, 14, 20, 29, 36, 37, 46–52, 59, 60
arithmetic, Babylonian 8; Egyptian 8
asteroids 74–5
astrolabe 26–8
astrology 19, 24, 31
astronomy, Indian 17, 28; Islamic 23–9; prehistoric 1–5
Averroes 28

B

Baghdad 23

Bede 31
Bellarmine, Robert 51
Bentley, Richard 92–3
Bessarion, Robert 35
Bessel F. W. 83, 84, 90, 91–2
Bode, J. E. 73
Bode's Law 73, 75
Boethius 29
Bologna University 32
Boulliau, Ismael 79
Bradley, James 82, 83, 84, 88–9
Brahe *see* Tycho
British Catalogue 72, 82–3
Bunsen, William 108
Buridan, Jean 34

C

Cairo Observatory 24
Calcidius 30
calendar, agricultural 4; Egyptian 6–7
Cassini, G. D. 66
cepheids 81
Ceres 74
Challis, James 75, 76
Chéseaux, J.-P. L. de 94–5
Christian IV (King) 45
chronometers 72–3
churches, orientations of 2–3
Cicero 30
Clairaut, A.-C. 70
comets 46, 47, 65, 67; Halley's Comet 69–70
Commentariolus 37
conjunction, planetary 43
Copernicus, Nicolaus 35–41, 42

Visit the
VERY SHORT INTRODUCTIONS
Web site

www.oup.co.uk/vsi

➤ **Information** about all published titles

➤ News of **forthcoming books**

➤ **Extracts** from the books, including titles not yet published

➤ **Reviews** and views

➤ **Links** to other **web sites** and main OUP web page

➤ Information about **VSIs in translation**

➤ **Contact** the editors

➤ **Order** other **VSIs** on-line

Expand your collection of
VERY SHORT INTRODUCTIONS